Environmental Footprints and Eco-design of Products and Processes

Series editor

Subramanian Senthilkannan Muthu, Hong Kong, Hong Kong SAR

More information about this series at http://www.springer.com/series/13340

Miguel Angel Gardetti
Subramanian Senthilkannan Muthu
Editors

Handbook of Sustainable Luxury Textiles and Fashion

Volume 2

 Springer

Editors
Miguel Angel Gardetti
Center for Studies on Sustainable Luxury
Buenos Aires
Argentina

Subramanian Senthilkannan Muthu
Eco-design Consultant
Global Sustainability Services
Hong Kong
Hong Kong SAR

ISSN 2345-7651   ISSN 2345-766X (electronic)
Environmental Footprints and Eco-design of Products and Processes
ISBN 978-981-10-1263-1   ISBN 978-981-287-742-0 (eBook)
DOI 10.1007/978-981-287-742-0

Springer Singapore Heidelberg New York Dordrecht London

Printed on acid-free paper

Springer Science+Business Media Singapore Pte Ltd. is part of Springer Science+Business Media (www.springer.com)

Preface

This book is Volume 2, which is the successor of our previous Volume 1 on Handbook of Sustainable luxury in Textiles and Fashion. This second volume comes out with eight important chapters to deal with the other new arenas of sustainable luxury in textiles and fashion, which were not covered in Volume 1.

Chapter "Loewe: Luxury and Sustainable Management" introduces Loewe—a company from the luxury industry and their sustainable management principles. This chapter deals with the model of sustainable value creation that integrates four elements: environment, innovation, stakeholder management, economic value and potential of growth.

Chapter "Long-Term Sustainable Sustainability in Luxury. Where Else?" deals with the past and current intersections of sustainable luxury and attempts to draw a conclusion on its future. This chapter includes interesting examples of luxury brands integrating themselves to sustainability and aims to put luxury and sustainability in the context of consumers and brands.

Chapter "Pineapple Leaf Fibre—A Sustainable Luxury and Industrial Textiles" revolves around the exploration of pineapple leaf fibre for sustainable luxury and industrial textiles applications. This chapter includes the discussions pertaining to the extraction methods of fibre and evaluation of fibre properties and importantly covers the sustainable utilization of the by-product generated during the process of fibre extraction.

Chapter "Beyond Appearances: The Hidden Meanings of Sustainable Luxury" deals with the dimensions that characterize sustainable luxury and the sources of the dialogue between luxury and sustainability. The important aspect is discussed with the aid of four Italian case studies.

Chapter "Irreplaceable Luxury Garments" addresses the value of emotional engagement and the state of irreplaceability of luxury garments, as one way of approaching the topic of sustainable luxury fashion. This chapter also presents the concept of sustainable luxury fashion the framework on the user's relationship to design products and also deals with the reframing luxury fashion through time and human presence.

Chapter "The Devil Buys (Fake) Prada: Luxury Consumption on the Continuum Between Sustainability and Counterfeits" deals with the determinants of the "dark side of luxury consumption", one of the largest challenges in luxury brand management: the increased demand for counterfeit branded product. An empirical investigation of a multidimensional framework of counterfeit risk perception and counterfeit shopping behaviour as perceived by distinct consumer segments is also discussed in this chapter to a greater extent.

Chapter "The Luxury of Sustainability: Examining Value-Based Drivers of Fair Trade Consumption" includes the discussions pertaining to the examination of the luxury of sustainability against the backdrop of the research questions concerning a proposed similarity of consumer associations between luxury and ethical products. This chapter presents the details on an empirical investigation of a multidimensional framework of intrapersonal fair trade orientation, fair-trade-oriented luxury perception and fair-trade-oriented customer perceived value with reference to the recommendation of fair trade products.

Chapter "The Sustainable Luxury Craft of Bespoke Tailoring and Its' Enduring Competitive Advantage" deals with how the luxury sector sustains its current consumption momentum and simultaneously espouses the logic of sustainable development. This chapter also deals with the following topics: concise chronology of cultural and societal events of humanity's affinity, slow fashion philosophy, the theory of the relational view and the importance of relational networks and the artisanal practices of the Savile Row tailors and their relational production network and retail facades.

We would like to take this wonderful opportunity to thank all the contributors of the chapters presented in this book for their intensive efforts in bringing out this second volume of this handbook so successfully with enriched technical content in their chapters. We are very much confident that this handbook will certainly become as an important reference for the researchers and students, industrialists and sustainability professionals working in the textiles and fashion sector.

Contents

Loewe: Luxury and Sustainable Management

Miguel Angel Gardetti

Abstract The sustainable luxury—the one that requires from organizations, a social and environmental performance of real excellence throughout the value chain—local "contents" based on culture should not be absent. Thus, sustainable luxury also means craftsmanship and innovation of the different nationalities, preserving the local cultural heritage. Loewe, an organization belonging to the LVMH group since 1996, is the leading brand of luxury leather in Spain and its name is related to the "masters of leather." At a global level, Loewe is in the luxury sector having started as a leather craftsman association in downtown Madrid in 1846. The development and manufacture of products offered by Loewe are carried out in Spain and shipped to the main markets in Europe, Japan, China, and the rest of Asia. Its products also include, beyond fragrances made at LVMH group, licensed sunglasses. Despite the fact that this brand was given a recognition as Lifetime Contribution to sustainable luxury Development within the framework of the IE Award for Sustainability in the Premium and Luxury Sectors on July 2, 2014, this case analyzes the company under the model of sustainable value creation developed by Professor Stuart L. Hart (Hart in Harvard Bus Rev 75:66–76, 1997; Hart in Capitalism at the crossroads. Wharton School Publishing, Upper Slade River, 2005; Hart in Capitalism at the crossroads—aligning business, earth, and humanity. Wharton School Publishing, Upper Slade River, 2007; Hart and Milstein in MIT Sloan Manage Rev 41:23–33, 1999; Hart and Milstein in Acad Manage Executive 17:56–67, 2003) that integrates four aspects: environment, innovation, stakeholder management, and potential for growth. To develop this case, the author collected background and information of the company from corporate documents (2013 Sustainability Report and the document "Traditional Values and Commitment to the Future" from 2012). This source was

M.A. Gardetti (✉)
Center for Studies on Sustainable Luxury, Av. San Isidro 4166 PB "A",
C1429ADP, Buenos Aires, Argentina
e-mail: mag@lujosustentable.org
URL: http://www.lujosustentable.org

© Springer Science+Business Media Singapore 2016
M.A. Gardetti and S.S. Muthu (eds.), *Handbook of Sustainable Luxury Textiles and Fashion*, Environmental Footprints and Eco-design of Products and Processes, DOI 10.1007/978-981-287-742-0_1

supplemented with interviews with members of the company's sustainability team. Loewe creates sustainable value through different sustainable practices that add value to the brand. They are energy efficiency, quality and environmental value products, supply chain management and dialog, diversity integration, acquisition of new capabilities for sustainability, and at strategic level, Loewe develops a sustainable vision going forward realizing its growth potential.

Keywords Luxury · Sustainable luxury · Loewe · Sustainable management · Sustainable value creation

1   Introduction

Sustainable development is a new paradigm, and this requires looking at things in a different way. And while luxury has always been important as a social determinant, it is currently starting to give place for people to express their deepest values. Thus, sustainable luxury promotes the return to the essence of luxury with its ancestral meaning, the thoughtful purchase, the artisan manufacturing, the beauty of materials in its broadest sense, and the respect for social and environmental issues (Gardetti 2011). Therefore, sustainable luxury will require—from those companies in this sector—a social and environmental performance of true excellence, which means starting with an internal change process, encouraging sustainable business practice across the organization and the supply chain. Consequently, first of all, this chapter introduces Loewe—a company from the luxury industry created by Enrique Loewe Roessberg in 1846, which currently belongs to the LVMH group. Then, it shows the model of sustainable value creation that integrates four elements: environment, innovation, stakeholder management, and potential of growth (Hart 1997, 2005, 2007; Hart and Milstein 1999, 2003), to end with an analysis of the company in light of the above model and a few conclusions.

2   Methodology

To develop this case, the author collected background and information of the company from corporate documents. Among them, I can mention the 2013 Sustainability Report, and the document "Traditional Values and Commitment to the Future" from 2012, addressing the current relationship between the company and sustainability. Moreover, this source was supplemented with interviews with members of the company's sustainability team.

3 Luxury and Sustainability

One of the most widely accepted definitions of sustainable development is the one proposed by the World Commission on Environment and Development (WCED 1987) report, Our Common Future, which defines sustainable development as the development model that allows us to meet present needs, without compromising the ability of future generations to meet their own needs. According to this report, the three pillars of sustainability would be "people, profit, and planet" (Bader 2008). Sustainable development is not only a new concept, but also a new paradigm, and this requires looking at things in a different way. It is a notion of the world deeply different from the one that dominates our current thinking and includes satisfying basic human needs such as justice, freedom, and dignity (Ehrenfeld 1999).

In turn, Coco Chanel once defined luxury as "(…) *a necessity that begins when necessity ends*" (Coco Chanel quoted in Okonkwo 2007: 7). In this same line, Heine (2011) defines luxury as something desirable and more than a necessity. These definitions depend on the cultural, economic, or regional contexts which transform luxury into an ambiguous concept (Low 2010). Luxury is a sign of prosperity, power, and social status since ancient times (Kapferer and Basten 2010). Christopher L. Berry in his work "The Idea of Luxury" from 1994 establishes that luxury has changed throughout time and that it reflects social norms and aspirations. This transformation of the idea of luxury was necessary to adapt to global changes (Shamina 2011). Luxury is the intimate perception and experience of things that turns them from unnecessary into indispensable. True elements of luxury rely on the search for beauty, refinement, innovation, purity, the well-made, what remains, the essence of things, the ultimate best (Girón 2012).

In the introduction of the book titled "Sustainable Luxury: Managing Social and Environmental Performance in Iconic Brands," Gardetti and Torres (2014) describe the evolution of the sustainability–luxury relationship between 2003 and 2011—a relationship that they first noticed in the book "Deluxe—How Luxury Lost its Luster," by Thomas (2007). Some of the aspects of this evolution are as follows:

The two most important conferences of this sector held in 2009 addressed these issues and focused their discussions on the assessment of these changes in the consumer and the new concept of success for achieving a "sustainable" luxury. One of them, organized by the International Herald Tribune in New Delhi (India), was called "Sustainable Luxury Conference." The other, promoted by Financial Times in Monaco with the attendance of Prince Albert, was titled "Business and Luxury Summit—Beyond Green: economics, ethics and enticement."
The book "Inside Luxury" written by Girón (2009) that presents a documented study of luxury and its future and how "sustainability" has influence on it.
The conduction by UNCTAD—United Nations Conference on Trade and Development—together with Green2Greener of the Conference "Redefining Sustainability in the International Agenda—Inspiring Greater Engagement in

Biodiversity Issues" with the participation of several luxury brands and the subsequent creation of the Responsible Ecosystem Source Platform (RESP) initiative. The creation—in early 2010—of the "Centre for Studies on Sustainable Luxury" whose mission is to assist companies in this sector in the transition toward sustainability, thus encouraging sustainable business practices across all areas of the organization and their supply chain. To this end, academic learning and research will become vital and current for future "sustainable" leaders.

The luxury/sustainability integration began to consolidate in 2011 through a series of events and writings, some of which were academic. Thus, it should be noted that in 2012, María Eugenia Girón led a team of experts in order to develop and publish the "Diccionario LID sobre Lujo y Responsabilidad" (LID Dictionary on Luxury and Accountability), featuring over 2000 definitions regarding the luxury sector. Moreover, in December 2013, Miguel Angel Gardetti and Ana Laura Torres (2013) published—as guest editors—a special issue on "sustainable luxury" within the framework of the Journal of Corporate Citizenship, and two books were published in 2014. The first one was called "Sustainable Luxury and Social Entrepreneurship: stories from the pioneers," whose editors were Miguel Angel Gardetti and María Eugenia Girón (2014), and a book called "Sustainable Luxury: Managing Social and Environmental Performance in Iconic Brands," by Miguel Angel Gardetti and Ana Laura Torres, was published in October 2014. The fourth edition of the IE Award for Sustainability in the Premium and Luxury Sectors was also held that year, 2014. The main purpose of this award is to acknowledge the culture and practice of sustainability in the premium and luxury sectors, and hence of their communication, in order to encourage "more sustainable" and, therefore, "more authentic" sectors.

Luxury—according to Kleanthous (2011)—is becoming less exclusive and less wasteful and more about helping people to express their deepest values.

Sustainable luxury is the returning to the essence of luxury with its ancestral meaning, to the thoughtful purchase, to the artisan manufacturing, to the beauty of materials in its broadest sense, and to the respect for social and environmental issues (Girón 2009).

Craftsmanship is a social construct that represents the cultural heritage of every region. Its expression, communication, and trade require specific channels and, of course, cultural protection for both the settlers and the consumers of these products on an international basis. Handicrafts imply the recognition and respect for one's own local characteristics and for the typical products that express and keep alive the culture of every region in the world.

Therefore, most artisan groups—including aboriginal artisans—wish to preserve the rooted local values as well as their beliefs in their relationships with the society and with the environment. In the global marketplace, handicrafts are being purchased by consumers who share those values, reject large-scale manufacture and mass production, and look for authentic "local" handmade objects (Grimes and Milgram 2000).

Sustainable luxury would not only be the vehicle for more respect for the environment and social development, but it will also be synonym for culture, art, and innovation of different nationalities, maintaining the legacy of local craftsmanship (Gardetti 2011). This aligns with the inherent features of craftsmanship: traditionalism, popular authenticity, manual prominence, individual domestic production, creative sense, aesthetic sense, and specific geographic location (FIDA, PRODERNOA and FLACSO 2005).

This shows a clear relationship between craftsmanship and sustainable luxury since—additionally and according to Girón (2012)—the true elements of (authentic) luxury rely on the search for beauty, refinement, innovation, purity, the well-made, what remains, the essence of things.

4 Creating Sustainable Value

The challenges associated with global sustainability, viewed through the appropriate set of business lenses, can help to identify strategies and practices that promote the creation of value. Therefore, the sustainable enterprise represents the potential for a new approach in bringing the private sector closer to development, including poverty, the respect for cultural diversity, and the preservation of ecological integrity (Hart 2005, 2007).

4.1 The Creation of Value

In this model, the creation of value—both short term and long term—is developed using two variables: a spatial variable and a temporal one.

The temporal variable reflects the firm's need to manage "today's" business, while simultaneously creating "tomorrow's" technologies and markets. Instead, the spatial variable reflects the firm's need to nurture and protect "internal" organizational skills, technologies, and capabilities, while simultaneously providing the firm with new perspectives and knowledge from "outside" stakeholders (Hart 1997, 2005, 2007; Hart and Milstein 2003).

The combination of these two variables—see Figs. 1 and 2—results in four different dimensions, crucial to the creation of value (Hart and Milstein 2003):

- Internal dimension and immediate term, such as cost and risk reduction.
- External dimension and immediate term (building of legitimacy).
- Future dimension (or long term) and internal (innovation and repositioning).
- Long-term dimension and external (credible expectations of growth).

To maximize value creation companies must act efficiently and simultaneously in the four dimensions.

Fig. 1 The value creation variables. *Source* Designed by the author (adapted from Hart and Milstein 2003)

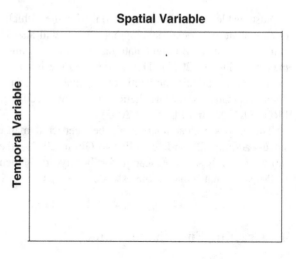

Spatial Variable

Temporal Variable

Fig. 2 Dimensions for value creation. *Source* Designed by the author (adapted from Hart and Milstein 2003)

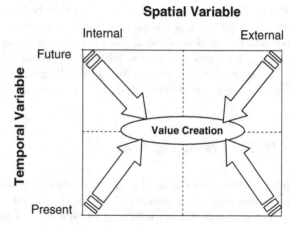

Spatial Variable

Internal   External

Future

Temporal Variable

Value Creation

Present

4.2 Global Drivers for Sustainability

According to Hart (1997, 2005, 2007) and Hart and Milstein (2003), there are four groups of drivers related to global sustainability that can be seen in Fig. 3 and which are explained below.

The first group corresponds to the growth of industrialization and its associated impacts, such as consumption of materials, pollution and waste and effluent generation. Thus, efficiency in the use of resources and pollution prevention are crucial to sustainable development.

A second group of drivers is associated with the proliferation and interconnection of civil society stakeholders, with high expectations for the companies' performance beyond their economic action. To achieve sustainable development, companies are challenged to operate in an open, responsible, and informed manner.

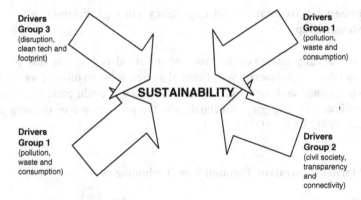

Drivers Group 3 (disruption, clean tech and footprint)

Drivers Group 1 (pollution, waste and consumption)

SUSTAINABILITY

Drivers Group 1 (pollution, waste and consumption)

Drivers Group 2 (civil society, transparency and connectivity)

Fig. 3 Sustainability drivers. *Source* Designed by the author

The third group of drivers regarding global sustainability is related to emerging technologies that would provide radical and "disturbing" solutions and that could make many of today's energy- and material-intensive industries obsolete. Thus, innovation and technological change constitute the keys to achieving sustainable development.

Finally, the fourth group of drivers is linked to population growth. In addition, the globalization of the economy affects the local autonomy, the culture, and the environment, causing a growing decline in developing countries (Hart 2005, 2007; Hart and Milstein 2003). A long-term vision that incorporates—to the traditional economic aspects—social and environmental aspects is essential for the achievement of sustainable development.

4.3 The Sustainable Value Structure: Connecting the Drivers with Strategies

Global sustainability is a complex multidimensional concept that cannot be addressed by any single corporate action. The creation of (sustainable) value implies that firms have to manage the four groups of drivers (Hart 1997, 2005, 2007; Hart and Milstein 2003). Each group of drivers has a strategy and practice, and these correspond to a particular dimension of value creation.

4.3.1 Growing Profits and Reducing Risk Through Pollution Prevention

Problems of raw materials consumption, generation of waste, and pollution associated with industrialization raise opportunities for firms to lower costs and risks, by developing skills and capabilities in ecoefficiency and pollution prevention (Hart 1997, 2005, 2007).

4.3.2 Enhancing Reputation and Legitimacy Through Product Stewardship

Product stewardship integrates the voice of stakeholders into business processes through an intensive interaction with external parties. It hence offers a way to lower both environmental and social impacts across the value chain and to enhance the firm's legitimacy by engaging stakeholders in the performance of ongoing operations (Hart 1997, 2005, 2007).

4.3.3 Market Innovation Through New Technologies

New technologies and sustainable technologies refer not to the incremental improvement associated with pollution prevention, but to innovations that leapfrog standard routines and knowledge (Hart and Milstein 1999). So rather than simply seeking to reduce the negative impacts of their operations, firms can strive to solve social and environmental problems through the internal development or acquisition of new capabilities that address the sustainability challenge directly (Hart 1997, 2005, 2007; Hart and Milstein 2003).

4.3.4 Crystallizing the Growth Path Through the Sustainability Vision

The vision of sustainability, which creates a map for tomorrow's businesses, provides members of the organization with the necessary guidance in terms of organizational priorities, technological development, resource allocation, and design of business models (Hart and Milstein 2003).

This model highlights the nature and the magnitude of those possibilities associated with sustainable development and relates them to the creation of value for the company. This is indicated in Fig. 4 which shows the strategies and practices associated with value creation of both short term and long term.

5 Loewe[1]

Loewe is the only Spanish luxury brand specialized in leather goods and silk items. It was founded in Madrid in 1846 and is owned by LVMH group since 1996. It is recognized worldwide for the quality of its crafted leather goods, the care with which it selects materials, as well as for its deep Spanish roots.

---

[1]This paragraph is mainly based on a company document called "Loewe, Madrid 1846" (unknown year). This source was supplemented with interviews with members of the company's sustainability team.

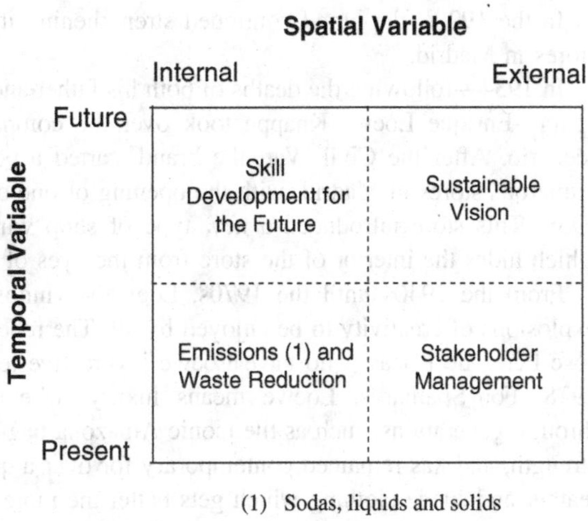

Fig. 4 Strategies and practices associated with the value creation of short term and long term and the essential elements for its development. *Source* Designed by the author (adapted from Hart and Milstein 2003)

Spatial Variable

	Internal	External
Future	Skill Development for the Future	Sustainable Vision
Present	Emissions (1) and Waste Reduction	Stakeholder Management

Temporal Variable

(1) Sodas, liquids and solids

Today, Loewe is a key player in 32 countries with over 160 stores worldwide. Over 1100 people work for the company, with headquarters in Madrid and subsidiaries in Hong Kong, Shanghai, and Tokyo.

5.1 The Origins... The Following Steps... and The Products (See Footnote 1)

Loewe's story goes back to 1846, when Madrid was getting ready for two royal marriages—that of her Majesty Isabel II of Bourbon to the Duke of Cadiz and that of Princess Maria Luisa Fernanda to the Duke of Montpensier. Also in that year, a group of Spanish craftsmen opened a leather goods workshop in the most commercial district in Madrid, Lobo Street (later Echegaray Street) in downtown Madrid. Then, in 1872, Enrique Roessberg Loewe, a German craftsman specialized in leather working, arrived in Madrid and decided to join forces with the leather workshop owners, and he established the brand. Twenty years later, in 1892, E. Loewe was present on the streets of Madrid, with a large poster staking his claim and marking an entire era. Afterward, the public, in general, became aware of the firm. Enrique Loewe brought his precision and technique, and Spain contributed with its sensuality, creativity, and peerless expertise in leather.

In 1905, Loewe became the "official supplier to the Spanish Royal Crown." When Enrique Loewe Hilton was at the head of the house of Loewe, King Alfonso XII gave the family business the title of Supplier to the Royal House.

The brand was consolidated with products that displayed quality and elegance, which led to very good economic results and business expansion in 1910 with the opening of the first store in Barcelona and then a second one also in Barcelona in 1918.

In the 1920s, the brand continued strengthening itself and began to open new stores in Madrid.

In 1934—following the deaths of both his father and grandfather just within five years—Enrique Loewe Knappe took over the company in an unstable political scenario. After the Civil War, the brand started a period of expansion, opening numerous stores in Madrid, with the opening of one on la Gran Vía Madrileña, in 1939. This store introduced a new type of shop window with a circular shape, which hides the interior of the store from the eyes of passersby.

From the 1940s until the 1970s, Loewe's window displays became glorious explosions of creativity to be enjoyed by all. The rebel soul creator back then was José Pérez de Rozas, who gave Loewe its creative lead from 1945 right up until 1978. For Spaniards, Loewe means luxury, whether about heirlooms passed through generations, such as the iconic Amazona bag—which name means female strength, and has remained contemporary for over a quarter of a century, or about leather and suede clothing which gets better the more it is worn. For men, Loewe spells leather jackets, attaché cases, and small leather goods. The Loewe silk scarves and ties in the rich and vibrant colors of Spain are unforgettable. Just as the sense of touch is vital to Loewe, so is the sense of smell. Scents for men and for women strive to bottle Spain's complex and magical allure.

Today, this is a global brand, part of the world's leading luxury group LVMH. The seductive touch of leather of unrivaled sheen, suppleness, and softness is now infused with modern glamor, thanks to Jonathan W. Anderson, the creative director who has brought his plugged-in fashion vibe to designs that are absolutely of their time and yet informed by heritage and craftsmanship.

5.2   Creativity, Craftsmanship: The Essence of Loewe (See Footnote 1)

Spain is internationally renowned for the quality of its lambskin and the very best comes to Loewe. "Cordero entrefino español" refers to lambs bred in the cool heights of the Spanish Pyrenees. Loewe's experts accept only the very small percentage of leather that can be judged perfect. As a result of this, Loewe's Napa has an unrivaled softness, suppleness, and sheen. That it is resilient yet sensual makes it superb for handbags and also for leather and suede garments, where the technical challenges can be even greater.

Then, there are the artisans whose craftsmanship is transmitted from generation to generation. Technical mastery is evident in leather bags, some of which are unlined, a rare feat which can only be achieved, thanks to the skills of those able to ensure that every stitch, inside and out, is perfect. There are few people in the world that can make leather and suede clothing to rival Loewe. A trench coat of good lightness signals an ongoing dedication to tradition fused with innovation.

Loewe will continue to focus on its core values of creativity and craftsmanship exemplified by the most sumptuous bags, leather and suede clothing for women and men, along with silken scarves and ties carrying tales of Spain.

5.3 The Sustainability Strategy[2]

Since 2012, the company has implemented a sustainability program based on six dimensions: design and development, purchases and supply chain, operations and logistics, sales and after-sales, in addition to commitment to employees and dialog with the stakeholders. As the goal of Lisa Montague, CEO and Managing Director of the company—is for "sustainability to become embedded in Loewe's DNA" (Loewe-Sustainability Report 2013: 11), each of the above dimensions is sponsored by a senior executive, who—together with Ms. Montague and the quality and sustainability team—make up the Sustainability Executive Committee.

By late 2013, the company had 1140 employees, with over 750 women, which is in line with the company's signature of the United Nations Women's Empowerment Principles. Additionally, it should be noted that 222 of the total employees are craftsmen. Based on the fact that Loewe is a traditional Spanish company, it features an intensive program to immerse in the cultural roots of the brand, which is held in Spain for foreign employees. As mentioned above, since craftsmanship is part of its essence, in February 2013, Loewe created a Leather Craft Training School in order to preserve the brand's characteristic know-how. Moreover, this initiative, which features 14 of Loewe's expert craftsmen as trainers, is conducted together with the Community of Madrid, the Town Council of Getafe —one of the Loewe plants is based in this city—and the European Commission's Social Fund. In 2015, external contractors will join the school to build a basis for dialog and know-how exchange.

The project called "la moda que nos va" (fashion that suits us)—created in 2013 and conducted jointly with FEAPS Madrid, an organization gathering 300 centers for people with intellectual disabilities and their families—allows employees to devote part of their work time to training these people in taking care of their image and feel self-confident in job interviews.

The Design Workshop—aimed at regarding sustainability as a source of inspiration and creativity, featuring external sustainable fashion experts—has a double approach: one which provides for "a bottom–top" reflection on product life cycle analysis and another one which offers "a top–down" view of product durability and repairability. In terms of durability, the purpose is to integrate it—within a collaborative process among different departments—into the earliest product

---

[2]This paragraph is mainly based on these two documents: "Loewe, Traditional Values and Commitment to the Future," from 2012, and "Loewe, Memoria de Sostenibilidad—2013" (Loewe 2012, 2013 Sustainability Report), from 2013. These sources were supplemented with interviews with members of the company's sustainability team.

development phase possible, checking for any potential production weaknesses and technical difficulties which might affect it. Moreover, the Leather Innovation Center is a space for dialog between the design and the raw material development teams, whose most salient feature is the leather environmental analysis through the development of a tool to assess the environmental impact of leather, developed together with the LVMH environmental department and the Canadian Research Centre—CIRAIG (Inter-University Research Centre for the Life Cycle of Products, Processes and Services). This tool helps, among others, to build dialog with suppliers about their strengths and weaknesses as to leather environmental aspects.

Every supplier adheres to a Suppliers' Code of Conduct that follows the same line as the LVMH group policies, as well as the guidelines of the International Labour Organization, the Universal Declaration of Human Rights, the Organisation for Economic Co-operation and Development (OECD), and the United Nations Global Compact conventions. While every supplier should also comply with the regulations on restricted substances and the animal welfare principles, Loewe co-funds improvement projects for its key suppliers' production processes. In addition to observing the animal welfare principles and the Convention on International Trade in Endangered Species of Wild Fauna and Flora (CITES), Loewe imposed some restrictions to the use of furs. For instance, it has strictly prohibited the use of Astrakhan, seal, and furs from trapping.

In terms of energy efficiency—between 2012 and 2013 and measured in kWh/m^2—the company made significant savings at its production plants (25 %), logistics center (17 %), and offices (17 %). However, store consumption increased in the same period of time and measured in kWh/m^2. Finally, the company features a waste management program to divide it into non-hazardous and hazardous waste (which includes containers with hazardous fluid remains and hazardous solid waste).

6 Creating Sustainable Value Loewe[3]: Analysis and Conclusions

The company reduces risks by managing its supply chain and conducting an analysis of its raw materials, both of which are strategic aspects for Loewe. Craftsmanship—also strategic for the company—involves trying to pass on skills, knowledge, and distinctive principles which—though unique—are constantly improved by Loewe, as it recreates how knowledge is usually passed on from one generation to the next in a family of craftsmen at its Leather Craft Training School. This handcrafted essence at Loewe can only but enrich the brand.

---

[3]The analysis shown in this paper has a limitation resulting from the use of Loewe documents and perspective without the perspective of their stakeholder.

Loewe avails of both the Design Center and the Leather Innovation Center to focus on innovation. "In-house," this means to develop the abilities, competences, and technologies that enable to continue positioning the firm for future growth as "masters of leather." Sustainable competences arising from research are the key to value creation. Companies that invest in innovative solutions tend to pursue novelty approaches to long-term challenges and create business environments consistent with the innovation process. This is about taking the lead to make incremental but radical improvements at product, process, and technology levels. This is what Rainey (2004) defined it as moving from thought to action.

Also, the company articulates a clear vision on its future and sustainability vision. This enables the company to trace a forward-looking "map," to set organizational priorities, to develop new products and technologies, and to allocate resources. This implies long-term leadership that promotes positive change (Rainey 2004).

Figure 5 depicts the strategies based on which the company builds sustainable value.

Organizational leadership and skill development are two main pillars to create sustainable value. This is about inspiring people inside the company to contribute to a more sustainable world. On the one hand, Ms. Montague—based on LVMH business vision—managed to align the organization behind her (sustainable) vision, and on the other, based on her commitment, she has inspired her team and consolidated a sustainable paradigm in the company.

Leadership also prevails company-wide innovation in new technologies—cold processes—since leadership also means to anticipate in the creation of solutions.

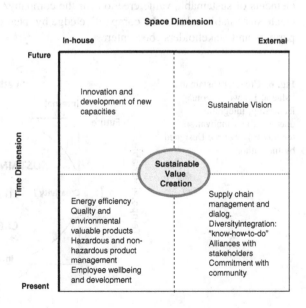

Fig. 5 Strategies and practices for creating sustainable value in Loewe's future. *Source* Designed by the author (adapted from Hart and Milstein 2003)

Sustainable value creation focuses on specific opportunities and challenges within a holistic management framework, responding to the business environment and creating sustainable positions for the future. And, here, the core is the integration of the company's capacities and resources with the capacities and mandates of the external business environment. This means that it binds all the internal and external resources and capacities to build unified management, embracing different perspectives, strategies, programs, processes, activities, and actions. The Leather Craft Training School, carried out jointly with the Community of Madrid, the Town Council of Getafe, and the European Commission's Social Fund, is a perfect example of that. This is what Rainey (2004) called integration and openness, as these two elements are essential for sustainable development, and through them, companies can build trust and legitimacy. Openness is essential to build reliable relationships with every stakeholder.

Even today, many companies do not recognize the strategic opportunities that result from sustainability. They are normally focused on, and invest their time in, short-term solutions, looking at the existing products and the different groups of stakeholders. However, Loewe develops sustainable ideas and new skills, and it builds trustworthy relationships with unconventional "partners," which enables it to create sustainable value. Within the framework of the Leather Innovation Center, this is the case of the alliance with LVMH environmental department and the Canadian Research Centre—CIRAIG (Inter-University Research Centre for the Life Cycle of Products, Processes and Services)—to develop a tool which helps, among others, to build dialog with suppliers about their strengths and weaknesses as to leather environmental aspects.

Figure 6 shows integration, openness, vision, and leadership as constituent elements of sustainable value creation for the company. All that enables Loewe to create sustainable value and a competitive edge by means of a model focused on its partners and stakeholders' best interest.

Fig. 6 Creating sustainable value at Loewe: vision, leadership, integration, openness, and innovation (creativity). *Source* Designed by the author

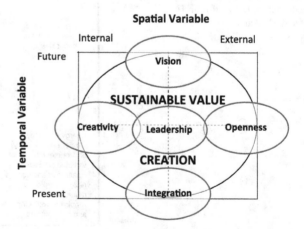

Acknowledgements The author would like to thank the members of Loewe for their valuable contribution to this case. Second of all, the author would like to thank Mrs. María Eugenia Girón— Director of IE Business School Premium and Prestige Business Observatory—for her contributions and continued reviews of this paper.

References

Bader P (2008) On the path to a culture of sustainability—conceptual approaches' Nachhaltigkiet. Retrieved from http://www.goethe.de/ges/umw/dos/nac/den/en3106180.htm. Accessed 7 Dec 2012

Berry CJ (1994) The idea of luxury—a conceptual and historical investigation. Cambridge University Press, New York

Ehrenfeld JR (1999) Cultural structure and the challenge of sustainability. In: Sexton K, Marcus AA, Easter KW, Burckhardt TD (eds) Better environmental decisions—strategies for governments, businesses, and communities. Island Press, Washington

FIDA, PRODERNOA, FLACSO (2005) El Sector de Artesanías en las Provincias del Noroeste Argentino. PRODERNOA, Buenos Aires

Gardetti MA (2011) Sustainable luxury in Latin America. conference delivered at the seminar sustainable luxury and design within the framework of the MBA of IE—Instituto de Empresa —Business School, Madrid, Spain

Gardetti MA, Girón ME (2014) Sustainability luxury and social entrepreneurship: stories from the pioneers. Greenleaf Publishing, Sheffield

Gardetti MA, Torres AL (2013) The journal of corporate citizenship a special issue on sustainable luxury (issue 52—Dec, 2013). Greenleaf Publishing, Sheffield

Gardetti MA, Torres AL (2014) Introduction. In: Gardetti MA, Torres AL (eds) Sustainability luxury: managing social and environmental performance in iconic brands. Greenleaf Publishing, Sheffield

Girón ME (2009) Inside Luxury. Editorial LID, Madrid

Girón ME (2012) Diccionario LID sobre Lujo y Responsabilidad. Editorial LID, Madrid

Grimes KM, Milgram BL (2000) Artisans and cooperatives—developing alternate trade for the global economy. The University of Arizona Press, Tucson

Hart SL (1997) Beyond greening—strategies for a sustainable world. Harvard Bus Rev 75(1):66–76

Hart SL (2005) Capitalism at the crossroads. Wharton School Publishing, Upper Slade River

Hart SL (2007) Capitalism at the crossroads—aligning business, earth, and humanity, 2nd edn. Wharton School Publishing, Upper Slade River

Hart SL, Milstein M (1999) Global sustainability and the creative destruction of industries. MIT Sloan Manage Rev 41(1):23–33

Hart SL, Milstein M (2003) Creating sustainable value. Acad Manage Executive 17(2):56–67

Heine K (2011) The concept of luxury brands. Technische Universität Berlin, Department of Marketing, Berlin

Kapferer JN, Basten V (2010) The luxury strategy—break the rules of marketing to build luxury brands. Kogan Page, London

Kleanthous A (2011) Simple the best is no longer simple. In: Raconteur on sustainable luxury, 3 July 2011. Source http://theraconteur.co.uk/category/sustainability/sustainable-luxury/. Accessed 7 Dec 2012

Loewe (2012) Loewe, traditional values and commitment to the future. Company document

Loewe (2013) Loewe, Memoria de Sostenibilidad—2013. Company document

Low T (2010) Sustainable luxury: a case of strange Bedfellows. University of Bedfordshire, Institute for Tourism Research, Bedfordshire

Okonkwo U (2007) Luxury fashion branding: trends, tactics, and techniques. Palgrave Macmillan, New York

Rainey DL (2004) Sustainable development and enterprise management: creating value through business.integration, innovation and leadership. Article presented at Oxford University at its colloquium on regulating sustainable development: adapting to globalization in the 21st century—8–13 August 2004

Shamina Y (2011) A review of the main concepts of luxury consumer behavior and contemporary meaning of luxury. Paper presented on the 2ème colloque franco-tchèque "Trends in International Business" Prague, 30 June 2011

Thomas D (2007) Deluxe—how luxury lost its luster. Penguin Books, New York

World Commission on Environment and Development—WCED (1987) Our common future. Oxford University Press, Oxford

Long-Term Sustainable Sustainability in Luxury. Where Else?

Coste-Manière Ivan, Ramchandani Mukta, Chhabra Sudeep and Cakmak Burak

Abstract We live in a world impacted by economic growth drivers of decreasing natural resources and increasing population. A shift toward a sustainable mindset does not seem a choice but a necessity. In today's extremely competitive luxury industry, sustainability seems like a paradox. The fashion and luxury business is producing more products, at a faster pace and change of seasons and constantly looking at decreasing costs. Many companies have moved production to Asia, especially to the more economically poorer nations. Is the demand of "sustainable luxury" enough for brands to consciously shift focus and change existing commercial production processes, do they realize the shift and awareness of the new customer who cares, and is it easy and practical to move toward "green" luxury? The purpose of this chapter is to explore the past and current intersections of

C.-M. Ivan (✉)
Luxury and Fashion Management SKEMA Business School, Sophia Antipolis, France
e-mail: ivan.costemaniere@skema.edu

C.-M. Ivan
Luxury and Fashion Management SKEMA Business School, Suzhou, China

C.-M. Ivan
Global Luxury Management SKEMA Business School, Raleigh, USA

C.-M. Ivan
Luxury Retail in LATAM, Florida International University, Miami, USA

R. Mukta
Neoma Business School, Reims, France

C. Sudeep
Armani Junior India, New Delhi, India

C. Sudeep
Unique Eye Luxury Apparels Pvt. Ltd., New Delhi, India

C. Burak
Corporate Responsibility—Swarovski, London, UK

C. Burak
School of Fashion, Parsons School for Design, New York, USA

C.-M. Ivan
North Carolina State University, Raleigh, USA

© Springer Science+Business Media Singapore 2016
M.A. Gardetti and S.S. Muthu (eds.), *Handbook of Sustainable Luxury Textiles and Fashion*, Environmental Footprints and Eco-design of Products and Processes, DOI 10.1007/978-981-287-742-0_2

sustainable luxury and draw a conclusion on its future. We take examples of luxury brands integrating themselves to sustainability. From understanding the production cycles, poor working conditions to how consumers are attracted to sustainable brands, we aim to help counter current challenges. Our methodology is comprised of qualitative data collected from various industry sources and interviews. The intent is to unravel the future of sustainable luxury with a lens on the manufacturing, packaging, and marketing efforts of a wide selection of luxury companies with different company turnovers. This chapter touches on a wide range of perspectives on the topic, including industry, designer, and retail, outlining a range of best practice strategies toward greater sustainability, while also acknowledging the complexity of the subject and its challenges, and the importance of the fashion and textile industries to livelihoods and business in general. The chapter specifically aims to put luxury and sustainability in the context of consumers and brands, understanding each through theoretical evidences from costly signaling theory, mimetic desire, pro-environmental values as well as practical phenomenon. Specific focus is given to the following: (1) Materials exclusive to luxury industry from exotic skins, high-quality tanned leather to furs, (2) Working conditions in manufacturing facilities, (3) Educating the customers, (4) Made in Europe versus developing countries, (5) Influence of technology in engaging with consumer on sustainability, (6) Evolution of business models that put transparency at the heart of emerging brands, (7) ROI, innovation, and luxury. The everlasting paradigm, (8) Emerging countries have also a great niche to boost, (9) Goodwill and know-how: sustainable by essence, (10) Branding the innovative sustainable way and new business models.

Keywords Sustainable luxury raw materials · Luxury and consumption · Engagement luxury consumer

1 Introduction

More and more luxury players are integrating corporate social responsibility and corporate environmental responsibility in their mission, objectives, strategy, and concrete actions (Cervellon and Shammas 2013), all under the pressure of being given low ranks on sustainability through nonprofit organizations (Bendell and Kleanthous 2007; Moore 2011). Increasingly, luxury consumers want to show that they understand and care about the environment and society. Luxury products are expected not only to look fabulous, but also to be environmentally and socially responsible. What satisfies luxury consumers, their perceptions of good quality and good design varies greatly across the world, and has been evolving over time. The global luxury industry has sometimes struggled to keep up, but has recently been challenged and rejuvenated by a new trend for authenticity and sustainability.

In this chapter we explore the following:

- Is there a strong and enriching relationship between consumer and product?
- Are luxury textile and fashion products being made that stimulate debate, that call for a deep sense of "meaning," or that necessitate the work of skilled craftspeople?
- Are luxury textiles and fashion products being designed that inspire confidence and ability that promote versatility, ingenuity, customization, or individual participation?
- Can emerging countries be the future of sustainable textile production?
- How craftsmanship in India is being preserved?
- What are the sustainable practices incorporated while manufacturing in India?

Each section elaborates on the key aspect of understanding sustainable luxury from consumers', retailers', and manufacturers' point of view. Directions for managerial practices are discussed and exemplified with brand cases.

2 Methodology

Through our chapter, we explain what sustainable luxury means in today's times. Going beyond the norm of simple theoretical explanations, we argue upon the theoretically defined concepts and managerial practices that need to be readdressed. This chapter is based on two main areas: First, various literature sources have been consulted, from academic journals, the luxury industry professional magazines, and a combination of sustainability and luxury. This also included several web sources such as blogs, online articles, and company Web sites. In addition to this collation and parsing of information, we have drawn upon our discussions with various couture designers, artisans, craftspeople, and opinion makers in the fashion and luxury business in India and across the world. Various reports on sustainability and luxury in the public domain have been referenced. Methodology is primarily qualitative based on the expertise of the contributors' long-term experience and existing relationships in the industry. Interviews are one of the key methods to collect up-to-date information from the key sustainability professionals in the luxury brands as well as NGOs that are working in textile and fashion industries.

As a result, we looked at the production cycle of luxury products, focusing on textiles and leather. Taking some examples and depending on their engagement in sustainability, we explored some of the remarkable works being done already. Finally, we analyzed the outcome seeking to identify whether sustainability is a challenge or an opportunity for the luxury sector.

3   Conceptual Discussion

3.1   Paradigm Shift from Luxury to Sustainable Luxury

As the alarm wakes up, Amy at her designer home in Shanghai and her House Keeper detail out her social engagements at breakfast, and she starts to get ready. While choosing her dress, she thinks about her girlfriends she will be meeting soon and remembers to wear her new "VogMask." She is not among some of her friends who have left Shanghai for the weekend because of the pollution. However, this is not going to stop her from meeting her friends for some window shopping at Nanjing road, the address for most of the luxury stand-alone boutiques.

This does not seem to be a prediction for the future. It is already prevailing in our present. Caring for the environment, using recyclable packaging, saving on the carbon footprint while traveling, and decreasing wastage in any form all seem to be the extreme end of the spectrum which seem to be so far away from the world of luxury.

His Royal Highness, the Prince of Wales participated through a video talk at the International Herald Tribune Luxury Conference in Miami (Sustainability, Artisanship and the Future of Fashion). Here he addressed that it is the collective responsibility of opinion leaders as well as senior members of the luxury fashion industry to secure a dignified future for our children. He stressed on focusing on the next Quarter Century rather than the next Business Quarter in the context of the luxury business being sustainable and viable. Loss of habitat and biodiversity, using unrenewable natural resources and not cultivating traditional techniques of crafts-manship, stands in our way of being sustainable. His message was both powerful and clear. The luxury industry needs to be responsible partners to the society and habitat of artisans that support the supply chain of luxury production. If not directly, he urged the industry leaders to support and look for sustaining the community that drives this very supply chain.

One could argue—How can it be luxury, if it is not excessive, ostentatious, over the top and created for the few because of its rarity and exclusivity? As it meets the eye, luxury and sustainability do not together. Do they? Or should they? In today's extremely competitive luxury industry, sustainability seems like a paradox. However, increasingly it is being talked about as not just the need of the hour, but the foundation on which true and unique luxury shall be built.

3.2   What Do We Mean by Sustainability?

As per EPA (United States Environmental Protection Agency), sustainability is based on a simple principle: Everything that we need for our survival and well-being depends, either directly or indirectly, on our natural environment. Sustainability creates and maintains the conditions under which humans and nature can exist in productive harmony that permit fulfilling the social, economic, and other requirements of the present and future generations. Sustainability is important to making sure that we have and will continue to have the water, materials, and resources to protect human health and our environment.

3.2.1 People, Planet, and Profit

When the triple bottom line (Slaper and Hall) accounting framework was introduced in the mid 1990s, it focused on the three dimensions of business—social, environmental, and financial. Also commonly called the 3 P's—People, Planet, and Profits, they are difficult to measure together as an index or a number. However, they provide a good lens to assess the focus of any business toward sustainability. When we look at sustainability in any organization through these three parameters, the typical focus is on Planet—taking care of our environment with regard to the production of goods. By nature, if we look at luxury products as those that have high quality as well as longevity because of the usage of rare raw materials, it could be argued that they are sustainable. This is because luxury is perceived to be rare and one of its kind. So customers buy few, but extremely beautiful and rare pieces rather than many that they may replace quickly. This is unlike the idea of the high consumption of fast fashion that keeps evolving very quickly, leading to a fast and large amount of production, leading to wastage.

3.2.2 Educating the Customer

How important it is to educate the customer about the sustainable practices of the luxury companies? And how do we achieve that? Research done by Griskevicius et al. (2010) shows that motives of gaining higher status can increase desire in consumers to buy greener products in public, even though the greener products might be more expensive than the non-green products. Thus, from a costly signaling theory perspective, buying sustainable luxury products in public can communicate pro-environmental behavior and altruism. This might be true in countries where status consumption is the norm, for example in India and China where mimetic desire and word of mouth drive the market of luxury industry (Ramchandani and Coste-Maniere 2012). But in a non-status consumption environment where show-off tendencies and ostentatiousness are not involved, the consumption paradigms are yet to be determined by researchers and experts in the industry.

Even though the previous research (Cervellon and Shammas 2013) indicates that sustainable luxury values in consumers encompasses sociocultural values (conspicuousness, belonging, and national identity), ego-centered values (guilt-free pleasures, health and youthfulness, hedonism, durable quality), and eco-centered values (doing good, not doing harm), but how do we influence the consumer values to consumption of sustainable brands? For this, we suggest that there is a real need to educate the consumers by various means of communication strategy, ecolabels and initiatives created to rank luxury brands on the scale of sustainable practices.

Therefore, we present in the followings section the initiatives and accreditations for sustainable luxury brands that are reinforcing new standards to define sustainable luxury.

What is Positive Luxury?

In 2011, Diana Verde Nieto and Karen Hanton founded Positive Luxury (About Us) a global award winning membership program revolutionizing the way consumers and brands get to know each other, and every brand featured on Positive luxury must take care on their sourcing of raw materials, manufacturing, and marketing services. Every brand featured on Positive Luxury takes great care with the sourcing of its raw materials, the manufacturing of its products, and the marketing of its services. Member brands to trust include Alexander McQueen, Balenciaga, Berluti, Boucheron, Burberry, Chaumet, Dior, DKNY, Emilio Pucci, Fendi, Gucci, Louis Vuitton, Marc Jacobs, Sergio Rossi, Hublot, and Valentino to name a few (Brands to Trust).

Green Carpet Challenge

Livia Firth is a name that needs no introduction when one thinks of sustainability and fashion and luxury. She is the Creative Director of Eco Age Limited and the driving force behind the Green Carpet Challenge® (GCC). The GCC Brandmark® has become one of the most sought after validations for sustainability in the global retail world. The GCC Brandmark is a guarantor of sustainable excellence and is awarded when the GCC social and ethical benchmark standards for a product or collection are met.

The first GCC Brandmark was awarded to Gucci (Gucci for Green Carpet Challenge) when it created, in partnership with Eco-Age, the world's first handbag collection made from zero-deforestation, certified Amazonian leather. Other brands, products, and initiatives that have been awarded the GCC Brandmark include the following:

- Chopard's Green Carpet Collections of High Jewellery.
- The London 2014 Stella McCartney Green Carpet Collection.
- The 2013 Green Carpet Capsule Collection, created by Christopher Bailey, Victoria Beckham, Christopher Kane, Erdem, and Roland Mouret.
- The Narciso Rodriguez (HEART) Collection for Bottletop.
- Handprint, the short film directed by Mary Nighy.

Though small steps, these initiatives aim to be conversation starters as well as triggers toward a larger awareness of the concept and more important the responsibility that all of us have toward a long-term sustainable future. As customers become more conscious and specific in their likes toward sustainable brands, they will have to follow ethical practices to become successful in the long term by choice.

3.3 Elucidating Typical Production Cycle

In this section, we examine the complete luxury goods' life cycle. To create a mood-board of a collection, raw material and sourcing are the first step toward sampling. Not only one has to look at the usage and application of raw materials, availability in the right quantities is imperative. And of course, costs have to be right. Major environmental footprint is caused by the clothing and textile sector, polluting around 200 t of water per ton of fabric (Nagurney and Yu 2012). Textile waste in Great Britain increased by about 2 million t per year between 2005 and 2010 (Kirsi and Lotta 2011).

Sustainable sourcing of raw material cannot be achieved as a quick fix unless an entire supporting ecosystem is created. This usually takes time, patience, and nurturing. Not many companies want to take ownership of this as it is a long-term commitment. Often, one is not even sure whether that particular raw material will be used often enough in each forthcoming collection, so why bother?

3.3.1 Sustainable Fabrics

Pier Luigi Loro Piana (the CEO and Deputy Chairman of the Loro Piana Group) was gifted by a Japanese friend a swath of beige fabric in 2009, and he had a hard time believing that the thread had been hand spun from the fibers of the lotus plant (Binkley 2010). The fibers of *nelumbo nucifera*, an aquatic perennial more commonly known as the lotus with its pink and white flowers, are sacred in parts of Asia. Through an artisanal handmade process, it takes approximately 32,000 lotus stems to make just 1.09 yards of fabric, approximately 120,000 for a costume.

It generally takes about 25 women making thread to produce enough yarn to accommodate one weaver. Keeping them moistened, the yarns are handwoven on looms into 100 yard (90 m) batches. This process takes approximately one month and a half to complete and also integrates a no-waste element as all parts of the lotus are utilized—using leftovers to make lotus teas, infusions, and flour. Best described as a cross between silk and linen, lotus flower fabric is naturally stain resistant, waterproof, and soft to the touch. This breathable, wrinkle-free fabric was once used to make robes for high-ranking Buddhist monks.

Despite the hurdles, the company has trademarked Loro Piana Lotus Flower fabric (Lotus Flower) and plans to sell lotus flower fabric from scarves to blazers that are priced upward of US$5500. Samatoa (natural fabrics) that produces lotus flower fabric also makes natural silk fabric, Kapok fabric, banana fabric, and organic cotton.

3.3.2 Good Versus Bad Pants

Jenny White and Verity White with Eco-Boudoir recently made a film MorethanPrettyKnickers (Pants Exposed) to raise awareness about fashion becoming a sustainable business driven by the demands of well-informed shoppers. The real aim is to educate consumers about the shocking facts of production of the most used and basic ingredient in the fashion business: cotton. It analyzes the impact of pesticides, labor condition toxic chemicals, and usage of water to present some startling facts. Their argument starts with a humble pair of pants. A pair of pants in the UK will create 19 kg of carbon dioxide equivalent (greenhouse gases) in its lifetime. So even if all 60 million people in UK are wearing a pair of pants right now—that is over 1 million t of carbon just today! Everyone knows the importance of reducing our carbon footprint and pants have no exception, and in textile and garment production, there is a lot of shipping and air freight involved to get the cotton from the fields to the knickers on your bottom. Using the analogy of "Bad Pants" and "Good Pants," they recommend the usage of fair trade cotton, Hemp and Silk fabrics, vegetable-tanned leather, bamboo, soy fabrics, and wool and request consumers to ensure products come from Fair Trade Practices.

3.4 Manufacturing in Developing Countries

In this section, we explore the pros and cons of manufacturing in developing countries and how to approach the challenges faced.

3.4.1 The Shirt on Your Back

A typical non-sustainable garment can make use of over 8000 toxic chemicals in textile creation processes (Hagen 2012). Besides the toxic chemicals used, the labor conditions are worsening in some of the developing countries. Not only they are prone to being exposed to these toxic chemicals but also work under .depriving situations. The Guardian captured one of the world's largest industrial tragedies in an eye-opening documentary called The Shirt on your Back (The Shirt on your Back). On 24th of April 2013, a nine-story factory called the Rana Plaza collapsed on the outer edges of Dhaka, Bangladesh. More than 1130 people died and more than twice were injured (2013 Savar Building Collapse). These were workers earning anywhere from £ 50 to £ 60 a month making clothes for western high-street fashion retailers including Mango, Matalan, and Benetton. Through this incident, many fashion industry manufacturers were criticized all over the world for their unsustainable practices in the Asian countries, which indicates that altruistic values in consumers define their concern for sustainable luxury practices. However, luxury sector may appear to exist outside of this system nurtured by the ideas of crafts-manship and design, but behind the glossy flamboyance lies the same dirt. The working conditions need to be addressed in order to not devaluate the respect of human lives. Luxury it seems has more respect for their merchandise than for people (Hoskins 2014).

3.4.2 Make in India Versus Made in India

An ancient Indian proverb says "We do not inherit the Earth from our ancestors, but rather borrow it form our children."

India the land of Yoga and Ayurveda has been synonymous with sustainability. In addition, in the field of sustainable luxury and fashion from sustainable production techniques, designing, raw materials, and supply chain, India is not far behind. This culturally diverse nation has opened up to the world by easing the FDI restrictions and welcoming companies to see India as a worthwhile destination for the production of goods and services.

In 2014, Make in India initiative was launched by the Government of India to encourage and attract international companies (Make in India 2014). The initiative aims at attracting international companies to manufacture in India. For the textile industry, India has been a lucrative country as it is the largest producer of natural raw materials such as jute and the second largest producer of cotton and silk. Launching a manufacturing unit in India for producing textiles is beneficial for sustaining a strong production base and catering to the increasing demand of the local and international consumers. Besides, the textile industry also employs directly over 45 million people in India, which serves as the largest employment generation sector for the country, consequently making the textiles and apparel industry estimated to reach US$100 billion by 2016–2017 from US$67 billion in 2013–2014.

The following section illustrates various sustainable aspects of luxury and fashion products made in India.

Leather

India ranks as one of the highest suppliers of leather in the world. Cities such as Agra, Kanpur, Chennai, and Kolkata are famous for exporting leather as furnished and semi-furnished to Europe and other parts of the world. India's export of leather and leather products reached US$3931.44 million (Council for Leather exports 2014). The leather industry in India thrived from the early nineteenth century to prepare shoes and leather goods for British army. The famous technique of vegetable tanning comes from the city of Kanpur. The technique involves tanning with tannin which is a class of polyphenol astringent chemical from the bark and leaves of many plants (Wikipedia 2014). This process of tanning has been used by various brands worldwide as a measure for producing sustainable leather and not causing skin allergies. Besides the process of tanning, the production of leather has also become an important subject for leather manufacturers in India. According to Zafar Iqbal Lari, a prominent manufacturer and exporter of leather from Kanpur, leather production process is scrutinized from beginning to the end of production for concerns of pollution. *"The process starts with cleaning the sledge with PET (Primary Effluent Treatment) plant. Which is segregated into chrome to weaken the strength of water and diluted into water at the level of PH 7. Which is then transferred to the All Tanneries Conveyance System and CETP (Common Effluent Treatment Plant) controlled by the government of India."*

Hidesign luxury leather brand from India

Dilip Kapur started his company—Hidesign, with a cobbler as a two-man workshop in 1978. Today, it has grown from its artisan roots to an international brand with over 60 exclusive retail stores and a distribution network across more than 20 countries. Products are all individually handcrafted using the finest leathers, many of which are purely vegetable-tanned using natural seeds and barks in their own tanneries. Different properties of the leather are checked such as the color (rich and natural) and the feel (oily and silky or dry and smooth), and then, the leather is cut by hand, numbered, and panel-matched so that each bag is sure to be made from the same batch (IIM 2006).

All leathers are full grain and have not been corrected with paint and pigment to hide natural defects. The natural and ecological tanning process enhances the intrinsic characteristics and individuality of their leathers and gives them a tremendous strength and durability. Each bag is therefore, by its nature, a limited edition. LVMH Group invested in his brand as early as 2007 and provides support (Hide and Chic).

Upon interviewing a sales representative Hina Manzoor at Hidesign store, Kanpur, we found that sustainability is of least concern to the customers. However, once they are told about the vegetable tanning and the benefits, they get more interested in the product. She mentioned that sustainability is still a challenge for the company as customers in India are not yet ready for it. Thus, emphasizing that communication of such an aspect is necessary in India to educate customers.

Luxury Ayurveda brand Forest Essentials

Forest Essentials was established in the year 2000 as a luxury Ayurveda company in India. This brand has been famous for producing skin care and hair products such as handmade oils, soaps from the Himalayas. The products are based on the recipes from Ayurveda after a long research. The highlight of the company's products is that each ingredient is handpicked from the forests in Himalayas and uses the therapeutic spring water from the region. From sustainable point of view, the products of Forest Essentials are never tested on animals, and it also engages local farmers to sustain local employability, does not involve child labors, and maintains strictly the finest traditional Ayurvedic recipes (About Us). The company has stores all over India. The estimated revenue of the company is Rs. 100 crore (1.6 billion USD), and in the year 2008, Estee Lauder bought 20 % stakes in the company (Rathore 2014). With higher involvement of Estee Lauder, Forest Essentials is set to launch stores globally as well.

Organic clothing brand No Nasties

No Nasties an organic cotton and fair trade clothing brand launched in 2010 has come a long way in promoting sustainability in India and abroad. The founders of the company started with the simple idea of establishing the brand with organic cotton derived from the Vidharbha belt in India and employing the local farmers (Janwalkar 2011). The cotton used in manufacturing has a certified clearance from the Global Organic Textile Standard. The founders aim to contribute in saving the planet through their line of clothes and employing local farmers in order to prevent farmer suicides (as the statistical rate of farmer suicides in India has been 1 in every 30 min). The price of cotton is set up by the farmers, no child labor involved and no middle men as the products are sold directly to the customer through the Web site of No Nasties.

Handicrafts

As Dr. Darlie Koshy points out (Koshy), handicrafts form the core of culture. They define cultural moorings and economic sustenance. While they use a wide range of inherent skills and local techniques, they provide a bridge to emotions and feelings through encoded values and aesthetics. These are the building blocks of the brand's DNA that cannot be duplicated. Crafts also form the best argument for sustainability. What better description of luxury than to wear handmade geography-specific products, made using rare material or techniques known to only a certain set of people. Crafts are true bespoke. To highlight the breadth and depth of craftsmanship available in textiles in India, Vogue India set upon a journey of exquisite handcrafted textiles. They then orchestrated collaboration with some of the biggest international labels to mold these textiles into something inspiring. Gucci created a dress from the famous "patola" fabric (a technique characterized by weaving of separately dyed warp and weft yarns to create surface motifs as per the design, usually in silk), Alberta Ferretti for Kanchipuram silk, Roberto Cavalli for Rajasthani Bandhini, Febdi for Bengali Jamdani, Christian Louboutin for Kanchipuram silk, Hermes for Bengali Kantha, Jimmy Choo for Benarsi brocade, Missoni for Lucknowi Chikankari, and many more (Project Rennaisance). This highlights the true possibility of creating luxury using traditional sustainable craftsmanship, which needs to be nurtured.

Shared Talent India

The Centre for Sustainable Fashion at the London College of Fashion started a project called Shared Talent India (Shared Talent India) to highlight the culturally vibrant and ecologically sensitive textile processing techniques that are present and being developed in India. It explores a variety of design opportunities and connects designers and buyers through sharing of knowledge and experience. India is perhaps the only country where story telling is practiced through textiles. Known as a sourcing hub for every major high fashion and luxury brand, it constantly amazes and surprises buyers.

Sustainable practices in the luxury and fashion industry are continuing to grow. Countries like India are upcoming and establishing their names for specialized sectors such as textiles, handicrafts, and leather as discussed in the earlier sections.

However, it is still yet to be explored how making products in developing countries will distinguish between sustainable practice paradigms for luxury companies benefitting and their country of origin effect like Made in Europe versus Made in Asia? Will there be a balance of sustainable standards in the production of textiles/goods and the quality luxury products have sustained so far with?

4   Can Sustainability and Luxury Go Together?

Luxury by definition is exclusive, rare, and hence long lasting. In some categories like watches, it is considered an honor to be passed as a family heirloom. The concept of sustainability can easily be integrated as part of the DNA of true luxury. It is the responsibility of the luxury industry to redefine itself so that their products embody this DNA of sustainability. The industry is also changing to believe that sustainability does not matter.

Since 2011, various luxury companies started waking up to the integration of sustainability in their DNA. This had some obvious advantages.

1. It is a part of the luxury business ethos—Jean Noel Kapferer, renowned French marketing Professor, observed the relationship as "luxury is at its essence very close to sustainable preoccupations because it is nourished by rarity and beauty and thus has an interest in preserving them. The unique values of the luxury business—Uniqueness, Timelessness and Heritage, all overlap with the ideology of Sustainability."
2. It is seen as an ethical business practice—The convergence of media high-lighting various aspects of the luxury business, whether it is outsourcing to developing countries, environmental impact, abusive employment policies, labor working conditions, or the health impact of the toxic residues present in food, textile, and cosmetics brings a high level of awareness among consumers. A luxury business aligned to the principles of sustainability is seen as a value enabler rather than arrogant. This creates a sense of belonging and awareness that engages the aware and involved luxury consumer. For example, Hermes invested in Shang Xia, a Chinese premium luxury brand of graceful, contem-porary handcrafted products. The usage of Cashmere Felt, Zitan Wood, Eggshell Porcelain, and Bamboo Weaving are some of the crafts revived and used as the brand story (Shang Xia—Chinese Fine Living). BMW's Efficient Dynamics technology was created to reduce harmful emissions and fuel consumption without sacrificing the comfort and pleasure of driving.
3. It is a clear differentiator—By stressing on the fundamental values of luxury, sustainability demarcates itself from the ever-changing, fast high fashion brands who promote consumption for the sake of owning such high fashion, rather than appreciating the value of rarity, the usage of noble materials, and craftsmanship.
4. Sustainability provides for a long-term return on investment—Luxury brands aligned and seen to be following sustainable practices echo their values of

timelessness and longevity. These values stand the test of time through generations and position the company being true to their core principles of craftsmanship. This enduring spirit weaves a certain longevity and story creating a long-term ROI.

5. Sustainability should be seen as a responsibility of luxury companies, in not so much a manner that it is required, rather than that luxury companies with their deep pockets and profits can contribute far easily to people, planet, and finally profits. While we can argue that the consumer shall eventually start distinguishing between environmentally conscious and sustainable organizations, it is the duty of luxury companies to become sustainable in whatever form and fashion quickly.

6. Sustainability is an opportunity for innovation. More and more designers are looking a creative ways in terms of material, design, packaging as well as giving back to society as a long-term value. To do this, they are seeking innovative ways to create what they have been creating for the last century or so. The Diffusion of Innovations Theory centers at creating conditions which increase the likelihood that a new idea, product, or practice will be adopted by members of a given culture. Diffusion of Innovation Theory predicts that media as well as interpersonal contacts provide information and influence opinion and judgment. Studying how innovation occurs, Rogers (1995) argued that it consists of four stages: invention, diffusion (or communication) through the social system, time, and consequences. The information flows through networks. The nature of networks and the roles opinion leaders play in them determine the likelihood that the innovation will be adopted. Ideas and hence practice of sustainability require early adopters to use the communication channels of the luxury business over time to be able to bring change to the social system.

5 Some Examples of Luxury Brands Integrated with Sustainability

According to Cervellon and Shammas (2013), luxury brands these days are complimentary with sustainability through ethos (e.g., eco-brands such as Stella McCartney (Kering/PPR) or Edun (LVMH) in luxury fashion or Tesla Roadsters and Venturi luxury cars), commitment all along the supply chain (Gucci group, LVMH, Porsche) and/or introduction of eco-collection and eco-lines (BMWi3 electric cars, Vranken-Pommery Pop Earth Champagne, Issey Miyake 132.5 origami design collection, Gucci sunglasses made out of liquid wood produced from sustainably managed forests).

In the following section, we show detailed cases of some luxury brands that have adopted sustainability as an integral part of the company.

5.1   PPR to Kering

In 2013, PPR became KERING. Their press release (PPR becomes KERING) mentioned—A world leader in apparel and accessories, Kering develops an ensemble of powerful Luxury and Sport and Lifestyle brands: Gucci, Bottega Veneta, Saint Laurent, Alexander McQueen, McQ, Balenciaga, Brioni, Christopher Kane, Stella McCartney, Sergio Rossi, Boucheron, Girard-Perregaux, JeanRichard, Qeelin, Puma, Volcom, Cobra, Electric and Tretorn. By 'empowering imagination' in the fullest sense, Kering encourages its brands to reach their potential, in the most sustainable manner.

The new corporate name was accompanied by an owl logo and a tagline, "Empowering Imagination." Its pronunciation, "caring," is meant to signify the group's approach of cultivating its brands. Kering recently announced a five-year partnership with the Centre for Sustainable Fashion at London College of Fashion (Kering). The project will involve annual talks for students, training, and internships at Kering brand's including Stella McCartney and a joint curriculum module focusing on the use of ethical materials and processes within the fashion supply chain. It is thought that recipients of the Kering Award for Sustainable Fashion will also receive financial support during their studies.

Sophie Doran writes on Gucci's steps toward sustainability (Doran)—Kering is arguably a leader when it comes to conglomerate-level corporate social responsibility in the luxury sector. As part of the group's overall profit and loss account, Kering's multi-tiered action plan focuses on the reduction of carbon dioxide emissions, waste, and water; sourcing of raw materials; hazardous chemicals and materials; and paper and packaging. Kering also aims to eliminate all hazardous chemicals from its production by 2020.

Leather for brands, such as Bottega Veneta, Sergio Rossi, and Alexander McQueen, will be 100 % sourced from "responsible and verified sources that do not result in converting sensitive ecosystems into grazing lands or agricultural lands for food production for livestock." When it comes to precious skins and furs—a long time hallmark of the Gucci brand—its goal is for 100 % to come from "verified captive-breeding operations or from wild, sustainably managed populations," where suppliers employ "accepted animal welfare practices and humane treatment in sourcing."

5.2   EDUN

(NUDE spelt backwards to stand for Natural) is a global fashion brand founded by Ali Hewson and Bono in 2005, to promote trade in Africa by sourcing production throughout the continent (EDUN).

In 2009, EDUN became part of the LVMH group. LVMH provides essential support to fulfill this vision. EDUN supports sustainable growth opportunities by supporting manufacturers, infrastructure, and community-building initiatives in Africa. It is helping to increase the trade throughout the continent with their apparel and accessories business. Currently, they manufacture more than 95 % of the collection in Africa. Over the past few years, they have collaborated with Conservation Cotton Initiative Uganda (CCIU) which assists by providing funding and training to displaced cotton farmers to help rebuild their businesses in northern Uganda. In fact, in a two-season partnership with Diesel, they created two collections inspired by the African desert produced entirely in Africa (Diesel+EDUN)

All of their clothing is made from organic fair trade materials using non-toxic biodegradable materials and organic water-based ink, even though this comes at an extra cost. They are a fair wage company hoping to alleviate poverty and provide equal opportunity.

5.3 Maiyet

Paul Van Zyl, Founder and CEO of **Maiyet** (Maiyet), pioneered a new luxury by creating a fashion brand that celebrates rare artisanal skills from unexpected places.

Maiyet's uniquely inspired, design-driven collection seeks to revive ancient techniques and elevate the next generation of master craftsmen from places such as India, Indonesia, Italy, Kenya, Mongolia, and Peru. The invention of the Jacquard loom in 1801 revolutionized the textile industry. It created a programmable system allowing elaborate complicated designs to be woven directly into fabric's construction. To create a jacquard textile, the desired pattern is coded onto a punch card that determines the placement of "warp" threads, strung vertically onto a loom. The loom operator conveys the horizontal "weft" thread through the warp on a "shuttle." With each subsequent stroke, the delicate jacquard pattern emerges from the loom's threads. Maiyet collaborates with master weavers in Varanasi who preserve the tradition of the hand-operated Jacquard loom to create finely wrought, original silks.

Hand-painted textiles are sketched directly onto textiles by master artisanal, which is then filled with details. Kalamkari—the Indian hand-painting tradition or "pen craft," was used in ancient times to decorate temple hangings with scenes and motifs from Hindu mythology. Once an art form widely practiced, hand painting is now a heritage craft disseminated by master artisans through years of apprenticeship (hand-painted scarf).

Maiyet's fine jewelry collection is hand crafted by artisan partners in France, India, and Italy and showcases rare stones, unique cuts, and a rich local heritage of fine jewelry craftsmanship. Subtle pave diamonds line the interior of a hand-carved horn and gold bracelet, an unexpected detail that nods to the considered design and impeccable craftsmanship of Maiyet's artisans. Diamonds range from full cut to polki slices, which are naturally earth-mined and uncut diamonds celebrated for their unique faceting and raw looking quality, and stones are set in gold as well as carved horn and bone pieces (Organic Geometric Necklace).

Maiyet is deeply committed to forging partnerships with artisans globally and has entered into a strategic partnership with Nest (Nest) an independent nonprofit organization dedicated to training and developing artisan businesses to promote entrepreneurship, prosperity, and dignity in places that need it most. Whether it is reviving a 500-year-old silk weaving tradition in Varanasi in India, or rescuing the Javanese art of Batik Textiles in West Java, Indonesia, and many more initiatives in Colorado, USA, Mexico, and Kenya to name a few. Nest is partnering with the world's most promising artisans to build sustainable businesses within the competitive landscape of today's global economy.

According to Van Zyl, the artisans with which Maiyet works face a number of specific challenges. "They lack design direction, access to markets, fair financing, the sort of training and rigour required for them to perform at the highest levels of the luxury market," he said. "We try through our model to offset all of these obstacles, so these craftsmen can turn their skills into viable businesses." (Business of Fashion Maiyet).

6 Conclusion and Crystal Ball Gazing

Many consumers believe that luxury brands adopting sustainable practices are purely commercial in nature (Achabou and Dekhili 2013). For consumers, ethical concerns can help to improve opinion and self-perception; they constitute an increasingly decisive factor in the psychological satisfaction afforded by luxury goods (Olorenshaw 2011). The initiatives of sustainable luxury practices are attractive to the consumers with altruistic values and pro-environmental behaviors. But it is still far from establishing a reputation of a sustainable brand. The transition of parent luxury brands to sustainable labels is not well defined, and further strategies need to be involved for educating the consumers.

We need to move from a world of "fast fashion" to a world of "slow fashion" where the dresses or handbags are designed and produced for a longer life cycle than merely being thrown away from one season to another. This is a beautiful echo of the brand DNA of luxury products and hence cements the idea of sustainable slow fashion in luxury. The nature of the slow fashion makes it luxury.

Raw material sourcing and processing has been addressed in this chapter along with future directions. We suggest brands to critically analyze the need of educating the consumers about sustainable practices adopted, in addition to imbibe research on identifying various customer segmentation for sustainable goods, for example status-driven consumption, altruistic consumption, and hedonic and utilitarian consumption. In a nutshell, sustainable luxury is intertwined with manufacturing practices adopted by companies and consumer behavior. Neglecting one aspect or the other cannot satisfy the sustainable luxury definition.

References

About Us (n.d.) Sustainable luxury forum. http://www.sustainableluxury.ch/about-us2/about-us/. Accessed Dec 2014

About Us (n.d.) Positive luxury. https://www.positiveluxury.com/about/. Accessed Dec 2014

About Us (n.d.) http://www.forestessentialsindia.com/about-us.html. Accessed 7 Jan 2015

Achabou MA, Dekhili S (2013) Luxury and sustainable development: is there a match? J Bus Res 66(2013):1896–1903

Animal Sourcing Principles (n.d.) Sustainable luxury working group. http://www.bsr.org/files/SLWG_Animal_Sourcing_Principles.pdf. Accessed Dec 2014

Bendell J, Kleanthous A (2007) Deeper luxury. www.wwf.org.uk/deeperluxury. Accessed Dec 2014

Binkley C (2010) New luxury frontier: a $5600 lotus jacket why Loro Piana is selling a $5600 Jacket painstakingly woven from a plant in Myanmar. http://www.wsj.com/articles/SB10001424052748703506904575592441000440092. Accessed Dec 2014

Brands to Trust (n.d.) Positive luxury. https://www.positiveluxury.com/brands/fashion/?&view=index. Accessed Dec 2014

Cervellon MC, Shammas L (2013) The value of sustainable luxury in mature markets a customer-based approach. J Corp Citizensh 52(12):90–101

Council for Leather Exports (2014) Analysis of export performance of leather and leather products. http://www.leatherindia.org/exports/exp-analysis-for-the-per-apr-oct-14.asp. Accessed 5 Jan 2015

Diffusion of Innovations (n.d.) Wikipedia.org. http://en.wikipedia.org/wiki/Diffusion_of_innovations. Accessed Dec 2014

Diffusion of Innovations Theory (n.d.) University of Twente. http://www.utwente.nl/cw/theorieenoverzicht/Theory%20clusters/Communication%20and%20Information%20Technology/Diffusion_of_Innovations_Theory/. Accessed Dec 2014

Doran S (n.d.) Gucci steps towards sustainability. Luxury society. http://luxurysociety.com/articles/2014/01/gucci-steps-towards-sustainability. Accessed Dec 2014

Eden Diodati (n.d.) Eden Diodati. http://www.edendiodati.com/philosophy. Accessed Dec 2014

EDUN (n.d.) http://edun.com/pages/about. Accessed Dec 2014

GCC Brandmark (n.d.) www.eco-age.com: http://eco-age.com/gcc-brandmark/. Accessed Dec 2014

Griskevicius V, Tybur JM, Van den Bergh B (2010) Going green to be seen. J Pers Soc Psychol 98 (3):392–404. doi:10.1037/a0017346

Green Carpet Challenge (n.d.) Vogue.com. http://www.vogue.com/1446381/the-green-carpet-challenge-2014/. Accessed Dec 2014

Hagen C (2012) Can sustainability and luxury go together? http://www.fastcoexist.com/1680581/can-sustainable-and-luxury-ever-go-together. Accessed Dec 2014

Hand Painted Scarf (n.d.) Maiyet.com. http://maiyet.com/shop/hand-painted-scarf–light-pinkchocolatered. Accessed Dec 2014

Indian Institute of Management, Ahmedabad (2006) Case study-hidesign—brand by design

Janwalkar M (2011) No compromises. http://archive.indianexpress.com/news/no-compromises/818577/2. Accessed 9 Jan 2015

Kapferer JN (n.d.) Sustainable future of luxury.theluxurystrategy.com. http://www.theluxurystrategy.com/site/wp-content/uploads/2012/05/What-do-clients-think-of-the-sustainable-future-of-luxury.pdf. Accessed Dec 2014

Kering (n.d.) Positve luxury blog. http://blog.positiveluxury.com/2014/11/kering-offer-sustainability-funding-london-college-fashion/. Accessed Dec 2014

Kirsi NA, Lotta HB (2011) Emerging design strategies in sustainable production and consumption of textiles and clothing. J Cleaner Prod 19:1876–1883

Kleanthous JB (2007) Deeper luxury WWF. http://www.wwf.org.uk/deeperluxury/. Accessed Dec 2014

Koshy DD (n.d) Handmade for sustainable fashion.darliekoshy.com. http://www.darliekoshy.com/docs/Handmade_for_Sustainable_Fashion_Futures-DrDarlieOKoshy.pdf. Accessed Dec 2014

Lotus Flower (n.d.) Loro Piana. http://www.loropiana.com/flash.html#/lang:en/product/FAE8554/6287. Accessed Dec 2014

Hoskins T (2014) Luxury brands—higher standards or higher markups? Guardian sustainable business. http://www.theguardian.com/sustainable-business/2014/dec/10/luxury-brands-behind-gloss-same-dirt-ethics-production. Accessed Dec 2014

Make in India (2014) Sector textiles. http://www.makeinindia.com/sector/textiles-garments/. Accessed Apr 2015

Maiyet (n.d.) Maiyet.com. http://maiyet.com/. Accessed Dec 2014

Moore B (2011) Style over substance report. Retrieved from Ethical Consumer Research Association Ltd at www.ethicalconsumer.org

Nagurney A, Yu M (2012) Sustainable fashion supply chain management under oligopolistic competition and brand differentiation. Int J Prod Econ 135:532–540

Natural Fabrics (n.d.) Samatoa. http://www.samatoa.com/PrestaShop/content/13-natural-fabrics-handmade-Cambodia. Accessed Dec 2014

Nest (n.d.) Build a nest. http://www.buildanest.org/. Accessed Dec 2014

Olorenshaw R (2011) Luxury and the recent economic crisis. Vie Sci Econ 188:72–90

Organic Geometric Necklace (n.d.) Maiyet.com. http://maiyet.com/shop/organic-geometric-necklace. Accessed Dec 2014

Pants Exposed (n.d.) MoreThanPrettyKnickers: http://www.morethanprettyknickers.com/. Accessed Dec 2014

PPR becomes KERING (n.d.) http://www.kering.com/en/press-releases/ppr_becomes_kering-0. Accessed Dec 2014

PPR to change Name (n.d.) Wall street journal. http://www.wsj.com/articles/SB10001424127887324103504578375903938246098. Accessed Dec 2014

Ramchandani M, Coste-Maniere I (2012) Asymmetry in multi-cultural luxury communication: a comparative analysis on luxury brand communication in India and China. J Glob Fashion Marketing 3–2(2012):89–97

Rathore V (2014) Estee lauder to up stake in forest essentials. http://articles.economictimes.indiatimes.com/2014-12-18/news/57195998_1_mira-kulkarni-forest-essentials-estee-lauder-group. Accessed on 8 Jan 2015

Rogers EM (1995) Diffusion of innovations, 4th edn. The Free Press, New York

Savar Building Collapse (2013) Wikipedia. http://en.wikipedia.org/wiki/2013_Savar_building_collapse. Accessed Dec 2014

Shang Xia—Chinese Fine Living (n.d.) Shang Xia. http://www.shang-xia.com/. Accessed Dec 2014

Shared Talent India (n.d.) Centre for sustainable fashion. http://www.sharedtalentindia.co.uk/home/. Accessed Dec 2014

The GCC Brandmark. (n.d.) www.eco-age.com: http://eco-age.com/. Accessed Dec 2014

The Shirt on your Back (n.d.)The guardian. http://www.theguardian.com/world/ng-interactive/2014/apr/bangladesh-shirt-on-your-back. Accessed Dec 2014

Slaper TF, Hall TJ (n.d.) The triple bottom line. Indiana business review. http://www.ibrc.indiana.edu/ibr/2011/spring/article2.html. Accessed Dec 2014

United States Environmental Protection Agency: 1 (n.d.). http://www.epa.gov/sustainability/basicinfo.htm. Accessed Dec 2014

Wikipedia (2014) Tanning. http://en.wikipedia.org/wiki/Tanning#Vegetable_tanning. Accessed 7 Jan 2015

Pineapple Leaf Fibre—A Sustainable Luxury and Industrial Textiles

Sanjoy Debnath

Abstract Pineapple (Ananas comosus) is a popular and one of the delicious fruit. Pineapple plant is mainly cultivated for its fruit. Since it is perennial crop, the plant is of no use after harvesting of the fruit. These pineapple leaves generate huge unutilized biomass and cause agricultural waste. In this approach, more emphasis has been provided to explore different possibilities of fibre extraction and convert it into value-added products. These products are mainly pineapple leaf fibre-based textiles for fashion and other important industrial applications. The textile industries involved mainly mini and cottage industries to develop these high-end pineapple leaf fibre-based textile products. The economics of the products depends on the innovative product design and newer applications of such pineapple leaf fibres. Overall, this chapter covers the sustainable utilization of the by-product generated during the process of fibre extraction. Potential uses and development of different industrial textiles, medical textiles, fashion textiles, pulp and paper, biofertilizer have been covered out of this novel sustainable pineapple leaf fibre.

Keywords Sustainable luxury raw materials · Luxury and consumption · Engagement luxury consumer

1 Introduction: Production, Yield, Luxury Fibre, Utilization, Sustainability, etc.

Pineapple is scientifically known as *Ananas comosus*. It is fruit-bearing plant of the family, Bromeliaceae, native to tropical and subtropical America but introduced other tropical places. This pineapple plant resembles the agave category. It has from

S. Debnath (✉)
Mechanical Processing Division, National Institute of Research on Jute & Allied Fibre Technology, ICAR, 12 Regent Park, Kolkata 700040, India
e-mail: sanjoydebnath@yahoo.com; sanjoydebnat@hotmail.com

© Springer Science+Business Media Singapore 2016   35
M.A. Gardetti and S.S. Muthu (eds.), *Handbook of Sustainable Luxury Textiles and Fashion*, Environmental Footprints and Eco-design of Products and Processes, DOI 10.1007/978-981-287-742-0_3

30 to 40 stiff, succulent leaves closely spaced in a rosette on a thick, fleshy stem. In general, it determinates inflorescence forms about 15 to 20 months after planting on a flower stalk 100–150 mm (4–6 in.) in length. The originally separate light purple flowers, together with their bracts, each attached to a central axis core, become fleshy and fuse to form the pineapple fruit, which ripens five to six months after flowering begins. The pineapple fruits of commercial varieties range between 1 and 2 kg in weight (Anonymous 2010).

History reveals that the Portuguese were apparently responsible for early dissemination of the pineapple. They introduced it to Saint Helena shortly after they discovered that island in 1502. Soon after, they carried it to Africa and, by about 1550, to India. Before the end of the sixteenth century, cultivation of the plant had spread over most of the tropical areas of the world, including some of the islands of the South Pacific (Anonymous 2015). It has been estimated that the total world production of pineapples ordinarily averages about 8,300,000 metric tons annually, majorly contributed by Thailand, Hawaiian Islands, Philippines, China, Brazil, Mexico, Côte d'Ivoire, India and Taiwan. After the harvesting of the crop (fruit), the plant is mostly considered as agricultural waste. Hamilton and Milgram (2008) covered textile aspects of pineapple leaf fibre emphasized more about the historical aspects and subsequent developments in pineapple leaf fibres in Asia-Pacific regions. Late 1970s Ghosh and Sinha (1977) understood and assessed the textile value of pineapple leaf fibre. They studied in details about the possibility of spinning and further weaving of pineapple leaf fibre in jute/flax spinning system to targeting fashion and technical textile material development. However, some efforts have been made in recent days to utilize its leaf extract for different commercial values. Pineapple leaf fibre is considered as fashionable textile fibre graded in between jute and cotton or jute and ramie. It has all textile properties and is capable of blending with jute, cotton, ramie and some other synthetic fibres (Ghosh and Sinha 1977; Ghosh and Dey 1988; Sinha 1982; Doraiswami and Chellamani 1993). So pineapple leaf fibre can capture an important position among natural fibres as potential commercial-grade textile fibre, but there is need of its assured supply to textile processing industry in sufficient quantities.

Upadhyay et al. (2010) and Dittrich et al. (1973) reviewed and concluded that most of the researchers have focused on the utilization of pineapple waste primarily for extraction of bromelain enzyme and secondarily as low-cost sustainable raw material for the production of ethanol, phenolic antioxidants, organic acids, biogas, organic fertilizer and fibre production. However, scientific and technological implications would produce better and more profitable markets for pineapple wastes utilization (Wang and Zhang 2009). Apart from the conventional use of pineapple leaf fibre in textiles, these fibres have immense potential to use as sustainable development of biocomposite material. As far as luxury material is concerned, pineapple leaf fibre extracted by hand/machine from the green leaves followed by suitable chemical/enzymatic treatment and further washed, bleached and dyed (if required), and finally dried lead to a high-textile-grade fibre. This process has been established in Philippines, further spinning and weaving produces a luxury textile product (pina fabric). Dey et al. (2009) focused different application dimensions of pineapple leaf fibre starting from fashion textiles to technical textiles and another aside of development of pulp and handmade paper. There are also evidences to develop reinforcement fabric from pineapple leaf yarn for rubber conveyor belt (Dey et al. 2009). Overall, this chapter will cover in details about the development of sustainable value-added fashion products from pineapple leaf as agro-waste.

2 Sustainable Fibre Extraction: Physiology of Pineapple Leaf and Its Fibre Presence, Different Methods of Fibre Extraction from Leaf, Their Process, Merits and Demerits

Pineapple leaf fibre is one of the abundant agro-waste sources (Fig. 1) that have been used for ages to be processed as different end product. Pineapple leaf fibre is a high-textile-grade commercial fibre, generally extracted by water retting. Pineapple leaf contains only 2.5–3.5 % fibre, covered by a hydrophobic waxy layer, which remains beneath the waxy layer (Paul et al. 1998; Banik et al. 2011). Out of the two main methods of fibre extraction from pineapple leaf are manual extraction (Fig. 2) and mechanical extraction. Uso and Dam (2013) concluded that comparisons between the machines are needed to ascertain the better function of machines. They also emphasized the importance where all natural fibres can be extracted using the same exact machine. Their data in this regard indicate that different types of methods used to extract the fibres though theoretically worked well for certain types of fibres but practically common fibre extractor does not perform well. The extracted fibres obtained though manual method owing to low fibre production but leads to fine quality of extracted fibres compared to machine extraction process (Kannojiya et al. 2013; Anonymous 2014). Out of different pineapple leaf fibre extractors available in India most popular are the extraction machine developed by National Institute of Research on Jute and Allied Fibre Technology (NIRJAFT), Kolkata (Fig. 3) and The South Indian Textile Research Association (SITRA).

Fig. 1 Pineapple leaves along with fruits

Fig. 2 Manual extraction of
pineapple leaf fibre
(Kannojiya et al. 2013)

Fig. 3 Mechanical extraction
of pineapple leaf fibre (Banik
et al. 2011)

Pineapple leaf contains only 2.5–3.5 % fibre, covered by a hydrophobic waxy layer (Kannojiya et al. 2013; Banik et al. 2011). Due to the presence of this hydrophobic waxy layer, fibre present in the leaf structure cannot be extracted simply by retting. During the retting or preferential rotting process, the microbes present in the retting liquor find difficulties to penetrate the inner core of the leaf because of this waxy coating. Hence, it has been found that either by simple manual extraction, i.e., scrapping by some sharp scrapper followed by retting/washing/chemical treatment produces a good quality of pineapple leaf fibres (Paul et al. 1998; Uso and Dam 2013; Kannojiya et al. 2013; Banik et al. 2011). Nag and Debnath (2007) have designed and developed a pineapple leaf fibre extraction machine where the machine can be operated with a ½ horsepower electric motor or an equal-power diesel-operated engine. The machine is designed in such a way that, at a time four green leaves can be processed. It has four zones, viz. feeding zone, scrapping zone, splitting zone and delivery zone. In the feeding zone, leaves are fed on the conveyor belt, which carry forward the green leaves through a pair of rubber feeding rollers. The feeding roller controls the movement of the leaves and fed to the scrapping roller, rotates at very high speed and removes the waxy coating from the top surface of the leaves without disturbing the fibrous layer. However, due to the variation in thickness of the leaf widthwise as well as lengthwise, the scrapping is not uniform. Hence, some portions like crop end areas as well as the both edges of the leaf blade are found not have been scrapped. This causes uneven retting, in fact, the unscrapped portion's retting does not occur and further fibres cannot be extracted from those areas. To overcome this problem, after the scrapping roller a pair of rotating serrated rollers have been used through which the scrapped leaves are passed. These serrated rollers split the leave blade lengthwise irrespective to its thickness as well as length and width of the leaves. Further the serrated leaves are collected in the delivery leaf-collecting bin. These scrapped and split leaves are collected together and made a bundle of approximate of 5 kg and allowed to retting for 5–6 days. The through washing of the retted fibres produces good lustrous pineapple leaves fibres. Table 1 shows the comparative pre-retting process on retting duration and corresponding fibre properties (Banik et al. 2011).

It has also been found that pineapple leaf fibres mostly present towards bottom surface of the pineapple leaves. However, most of these fibres lie below 1–1.3 mm below the top surface of the leaves (Banik et al. 2011). This thickness of the coating also depends on various factors and some of such important factors are the variety cultivated, age of the leaf, position of the leaf (side/centre of the leaf blade), etc. After the extraction of fibre from either manual or mechanical process, the fibres are mostly contained some natural gummy materials, which affects the fibre well separated during the spinning processes. Hence, some of the researchers held with concluded that some enzymatic degumming of pineapple leaf fibre will produce fine and good-quality pineapple leave fibres. Both the theoretical and practical aspects of enzymatic degumming of pineapple leaf fibre were explored by them. In their study, they purified pectase with high enzymatic activity from a strain screened by them was used. It has been found that the optimal result of degumming could be achieved when reaction conditions were pectinase dosage 8 %, pH value 7.0, at temperature 52 °C with treatment duration of 4 h. Post-treatment after degumming with xylanase for 45 min followed by treatment with H_2O_2 for 15 min would yield a good-textile-grade fibre for further textile utilization.

Table 1 Comparative pre-retting process on pineapple leaf fibre retting and fibre properties (Banik et al. 2011)

Experiment No.	Pre-retting procedure	Retting period (days)	Tenacity (g/tex)	Fineness (tex)
1	Manual combing	10	7.8	Variable
2	Machine scrapping (one side)	8	11.1	4.3
3	Machine scrapping (both side)	8	11.6	5.0
4	Machine scrapping plus combing (one side)	6	9.7	5.7
5	Machine scrapping plus maceration (one side)	6	16.7	3.4

3 By-product Utilization Generated During Fibre Extraction: Sustainable Utilization of the Agricultural By-product Wastes, their Disposal—Conversion from Waste Into Wealth

During any agricultural production, agricultural waste is one of the common problems. Pineapple fruit production also generated huge amount of agro-waste. Banik et al. (2011) found that the residual green sludge can be used for vermicomposting (complete within 45 days), which is rich in plant nutrients. Earthworm species African night crawler (*Eudrilus eugeniae*) has been used as inoculum for vermicomposting. The combined technology package for extraction of fibre and utilization of the residual biomass debris from the pineapple leaf scratcher machine for vermicomposting is economically viable for pineapple cultivators.

Sustainable utilization of the agricultural by-product wastes and economics
Banik et al. (2011) established that the green leaf waste generated during the pineapple fibre extraction from leaf is good source of compost. They maintained average moisture content in vermicompost cast was 50 % and the pH was 7.0. Their study also revealed that vermicompost contains more nitrogen (1.0–1.2 %), phosphorus (0.3–0.4 %) and potassium (0.4–0.5 %) almost similar C:N ratio than other compost, and hence, it is rich enough in NPK and will be suitable for agriculture. Earthworms are invertebrates and are of two types such as burrowing type and non-burrowing type. The non-burrowing types live in upper layer of soil surface and consume 10 % soil organic matter and 90 % added organic matter, whereas the burrowing type lives deep in soil. However, non-burrowing types are preferred which depend 90 % on soil organic matter and 10 % on added organic matter to make compost from pineapple leaf scrapping waste after removal of fibre. Moreover, earthworm inevitably consumes soil microbes during ingestion of the organic substrate and extracts nitrogen from microbes especially from fungi as suggested by Ranganathan and Parthasarathi (2000). This may be the reason for less number of fungi in vermicompost samples (Ranganathan and Vinotha 1998). It has been evidenced from the experiments of Banik et al. (2011) that 0.68 tons of vermicompost was formed from 1.5 tons of pineapple leaf scratching residues. One hectare pineapple cultivation land can produce 8 tons of fresh harvested agro-waste leaves after harvest of pineapple fruit. Present Indian market price per kg of the vermicompost and pineapple leaf is approximately INR 6/- and 25/-, respectively. Assuming average of 8 tons of pineapple leaf from one hectare of land and the fibre yield @ 2.5 % is 200 kg from one hectare only pineapple cultivation. Hence, the extra income from pineapple leaf fibre is INR 5000/- and INR 4080/- from vermicompost, i.e., INR 9080/- from one hectare of land from the integrated system of waste management. The payback period for the pineapple leaf scratcher machine is 4.5 years with break-even point 39.57 %.

4 Evaluation of Fibre Properties Extracted from Different Methods: Physical and Chemical Properties of Pineapple Leaf Fibres, Effect of Important Fibre Properties From Different Process of Extractions

Kannojiya (2013) reported that Pineapple leaf fibre could be modified by alkali, acetylation and graft copolymerization. Further, grafting process improved the thermal stability property of pineapple leaf fibre. The modified fibres showed that, with introduction of these treatments significant improvements on hydrophobicity, improved mechanical strength and chemical resistance obtained. Li et al. (2004) in their paper summarize the studies on chemical modification of pineapple leaf fibres and expound on the main chemical modification methods and subsequent results. The chemical composition of pineapple leaf has been compared with some plant fibres such as jute, ramie and cotton as shown in Table 2. The cellulose content of the cotton is highest followed by ramie, pineapple leaf fibre and jute fibre respectively, as found from Table 2. On the other side, the lignin content is 4.4 % which is much higher than cotton and ramie and lower than jute fibre. This is one of the reason the colour of the pineapple fibre is creamy white. The degree of polymerization of cellulose and crystallinity is close to jute fibre. Overall, the pineapple leaf fibre is somewhat better than jute fibre but inferior than cotton and ramie fibres. When look into some of the physical properties of pineapple fibres, it is clear that though the pineapple leaf fibre is little coarser than jute and ramie, but its extension at break is comparable to that of ramie and twofold that of jute fibres (Table 3). This gives better extensibility of the pineapple leaf fibre. Probable, lesser lignin content of pineapple leaf fibre (Table 2) is one of the responsible factors for comparatively higher extensibility in pineapple leaf fibre. An interesting part of this pineapple fibre is that low torsional rigidity compared to jute and ramie fibres. Since torsional rigidity of the pineapple leaf fibre is low, it requires less energy during the spinning

Table 2 Comparative chemical composition of pineapple leaf fibre with some important plant fibres in percentage (Banik et al. 2011)

Chemical constituents	Some important plant fibres			
	Pineapple leaf fibre	*Capsularis* jute	Ramie	Cotton
α-cellulose	69.5	61.0	86.9	94.0
Pentosans	17.8	15.9	3.9	0
Lignin	4.4	13.2	0.5	0
Fat and wax	3.3	0.9	0.3	0.6
Pectin	1.1	0	0	0.9
Nitrogenous matter	0.25	1.56	2.1	1.2[a]
Ash	0.9	0.5	1.1	1.2
DP of α-cellulose	1178	1150	5800	2020
Crystallinity of α-cellulose	57.5	55.0	70.0	68.0

DP Degree of polymerization;
[a]As protein

Table 3 Some important physical properties of pineapple leaf fibre compared with some important plant fibres (Banik et al. 2011)

Physical properties	Pineapple leaf fibre	Jute		Ramie
		Capsularis	Olitorius	
Fineness, tex	2.8	2.2	2.5	0.7
Tenacity, g/tex	26.1	25.0	23.9	45.0
Extension at break, %	3.0	1.5	1.5	3.5
Flexural rigidity, dynes cm^2	3.8	4.5	4.6	1.0
Torsional rigidity $\times 10^{10}$, dynes/cm^2	0.36	0.85	0.80	1.5
L/B ratio of ultimate cell	450	110	110	3500

of yarn out of this fibre. Overall, the pineapple leaf fibre is superior to jute fibre and inferior to ramie and cotton fibres.

Chakravarty et al. (1978) made an effort of study the tensile behaviour of pineapple leaf fibre in wet condition. A very interesting phenomenon has been observed in case of pineapple leaf fibre and yarn, when their tensile properties were studied in wet condition. The bundle strength of pineapple leaf fibre falls down by 50 % in wet condition, but the yarn strength in wet condition increases by about 13 %. This similar phenomenon has also been observed in case of jute fibre though both are multi-cellular lignocellulosic fibres but jute is bast fibre with mesh-like structure and pineapple is leaf fibre having non-mesh filament structure. Saha et al. (1990) studied the structural features and fracture morphology of raw and chemically treated pineapple leaf fibres using scanning electron microscopy. They clearly found the pineapple leaf fibres have a multicellular structure and alkali treatment reveals the ultimate fibres. They also found that the surface morphology progressively changes with gradual removal of non-cellulosic constituents such as lignin and hemicellulose. As far as the fracture morphology is concerned, from this study it is clearly evidenced that there is a uniform sharing of the load in the ultimate fibres of raw pineapple leaf. On the other side, the treated pineapple leaf fibres show irregular fractures characterized by the slippage of individual ultimate fibres. This study confirms that the lignin and hemicellulose present in pineapple leaf fibres are essential cementing materials that bind the ultimate cell components together. This causes them to break at approximately the same instant when a tensile load is applied. They concluded that during bleaching or alkali treatments of pineapple leaf fibres, necessary care should be remove sufficient amounts of lignin and hemicellulose as they act as binding material of the ultimate cells. Hence, this will prevent excessive slipping of the ultimate cells and help to retain good amount of tensile strength. Saha et al. (1991) investigated the infrared spectra of raw and chemically treated (NaOH and $NaClO_2$) pineapple leaf fibres using the KBr disc technique. Presence of bands due to lignin and hemicellulose in the spectra of the fibres and their changes due to chemical (NaOH and $NaClO_2$) treatment studied in details. Mukherjee and Satyanarayana (1986) have analysed the stress–strain curves for pineapple leaf fibre. They investigated that different parameters such as ultimate tensile strength (UTS), initial modulus (YM), average modulus (AM) and elongation of fibres have been calculated as functions of fibre diameter test length and test speed. UTS, YM and elongation lie in the range of 362–748 MN m^{-2}, 25–36 GN m^{-2}, and 2.0–2.8 %, respectively, for fibres of diameters ranging from 45 to 205 μm. UTS was found to decrease with increasing test lengths in the range 15–65 mm. Their study reveals that mechanical parameters have marginal changes with the change in speed of testing in the range of 1–50 mm min^{-1}. Further, they made an effort to explained results on the basis of structural variables of the fibre. Scanning electron microscope studies of the fibre reveal that the failure of the fibres is mainly due to large defect content both along the length and cross section of the fibre. The crack is always initiated by the defective cells and further aggravated by the weak bonding material between the cells (Mukherjee and Satyanarayana 1986).

5 Value Addition in Sustainable Developing Luxury Textiles and Industrial Applications: Sustainable Processing of Yarn/Fabric Development, Possibilities of Blending With Different Natural Fibres for Luxury Textiles, Designing Sustainable Textile Materials for Industrial Applications

In recent days, more attention has been paid to the use of renewable resources particularly from plant origin considering the ecological concerns of the planet Earth. On the other side, despite abundant availability from renewable resource of lingo-cellulose materials, very few attempts have been made about their proper utilization. This may be due to lack of availability sufficient structure/property data and awareness to the users. However, systematic studies may bridge this gap while leading to value addition to these natural fibrous materials.

Hamilton and Milgram (2008) go on a voyage through a variety of thought-provoking issues relating to marginal fibre artefacts across the remote areas of the Asia-Pacific region. They offered the anthropological approaches in their book on textile and apparels seeking broader perspectives in order to understand objects. In addition, they lavishly illustrated throughout, providing inspiration for anyone with an interest in the processes of fibre production as well as the woven textiles they crafted. Ghosh and Sinha (1977) and Sinha and Ghosh (1977) are pioneer in textile product development from pineapple leaf fibre. They used some special technique to spin pineapple in jute spinning machinery. In their study, they used 15 % oil emulsion followed by processing through softener machine and binned for 24 h. They used conventional jute breaker card followed by flax card to prepare the carded sliver instead of jute breaker card and jute finisher card. The optimum twist factor was found between 24 and 27 units in yarns of 70–170 tex. However, in admixture with jute, 10–15 % of pineapple fibre will improve the performance of jute-blended yarn and fine jute–pineapple-blended yarn can be produced. Further, with these fine pineapple yarn and pineapple–jute-blended yarn, plain and twill woven cloths have been developed for fashion fabric development. These lightweight fashion fabrics further used to design fashion bag, curtain and furnishing fabrics. With these, they concluded that for sustainable development of fashion fabrics use of pineapple leaf fibre or jute–pineapple leaf fibre-blended products have immense potential. A significant amount of research made by Ghosh et al. (1982) on processing of pineapple leaf fibre in cotton machinery. Before processing in cotton spinning system, they studied and compared the cotton, jute and pineapple leaf fibre about their physical and mechanical properties. They found that the 100 % pineapple leaf fibre is not at all possible to spin into yarn in cotton spinning machinery. They tried with different proportion of pineapple leaf fibre, viz. 50, 33, 20 % with cotton. However, with blend of 50 % pineapple with cotton is quite promising to spin pineapple–cotton-blended yarn. Though the spinning performance is poor in cotton–pineapple blend but a huge amount of cotton can be saved and thereby value-added luxury product can be made out of this cotton–pineapple-blended yarn. They concluded that during blending pineapple with cotton, the yarn tenacity falls significantly while increasing the proportion of pineapple leaf fibre more than 50 % in the cotton blend. In the same area of blending of pineapple leaf fibre, researchers also made an attempt to study the performance of pineapple leaf fibre and acrylic fibre blending in jute spinning system (Ghosh et al. 1987; Dey et al. 2009). They studied the fibre properties of pineapple leaf fibre and acrylic fibres and compared the similarities and dissimilarities of these two fibres. Five different blends of pineapple leaf fibre and acrylic fibre have been tried, viz. 87:13; 67:33; 50:50; 33:67 and 13:87. From all these blends, fine yarn in the tune of 84 tex yarn with 7.5 twist/in. were spun in wet spinning process where the rove was passed through the temperature bath (80–100° C) before spinning. They also spun the same yarns through dry spinning process. They compared the dry and wet spinning process and found that in wet spinning the breaking stress is reduced, but the breaking strain has been improved by 6 times. The optimum blend composition has been found from their studies is 67:33 pineapple–acrylic-blended yarn. The wet spinning performance is much superior to in dry spinning method. Finally, they also concluded that there is good scope for sustainable development of fancy apparel products out of these pineapple–acrylic-blended yarns (Dey et al. 2009). Doraiswami and Chellamani (1993) reviewed various research papers and concluded that pineapple leaf fibre has vast potential to develop luxury textiles for sustainable development of natural fibre-based products. They covered the whole process starting from fibre extraction to the final yarn and fabric from pineapple leaf fibre and its blends. The eco-friendly in nature and by-product utilization point must be considered, and hence, the end product is costly.

Composite, an important area wherein there are lot of scopes to utilize the pineapple leaf fibre as reinforcement material component. Pineapple leaf fibre-based wood plastic composite (WPC) is a very promising and sustainable green material to achieve durability without using toxic chemicals as evidenced form the studies by Chaudhary et al. (2012). Pineapple leaf fibre-reinforced composite can replace glass fibres in fibre-reinforced plastics in some of the application areas with different technical properties (Mohamed et al. 2009; Saha et al. 1993; Luo and Netravali 1999; Uma Devi et al. 1997; Liu et al. 2006; Luo and Netravali 1999; Arib et al. 2006). Lopattananon et al. (2006) investigated the performance of pineapple leaf fibre–natural rubber composites, wherein they emphasized on the effect of fibre surface treatments on ultimate pineapple leaf fibre-based rubber composite. A series of work made by Liu et al. (2005) and Mishra et al. (2001, 2004) made an important research contribution on soy-based bioplastic as natural resign and pineapple leaf fibre as natural reinforcement fibre component. Both the fibre and reinforcing materials are natural, they termed as 'green' composites, which were manufactured using twin-screw extrusion and injection moulding. They also optimized different processing parameters for optimum properties. George et al. (1995) optimized the pineapple fibre length to achieve the best performance of the PALF–LDPE composites. Pineapple leaf fibre-reinforced phenol-formaldehyde (PF) composites have been studied thoroughly by Mangal et al. (2003). They established good agreement between theoretical and experimental results obtained.

As far as medical and pharmaceutical industry is concerned, in the present era, a lot of importance has been understood for the applicability of nanocellulose in pharmaceutical industry as well as valuable product in other different industries. It is one of the very costly materials. Recent work in this line reported by Cherial et al. (2010) is pioneer in this area. They used steam explosion process for successful extraction of cellulose nanofibrils from pineapple leaf fibres. Steam-coupled acid treatment on the pineapple leaf fibres is found to be more effective in the de-polymerization and defibrillation of the fibre to produce nanofibrils of these fibres. The chemical constituents of the different stages of pineapple fibres undergoing treatment were analysed according to the ASTM standards. There are evidences from the XRD analysis results of the crystallinity of the treated fibre particles. Further, Cherian et al. (2010), characterized of the fibres by SEM (Figs. 4 and 5), AFM and TEM, supports the evidence for the successful isolation and size of nanofibrils from pineapple leaf. The developed nanocellulose promises to be a very versatile material having the wide range of biomedical and biotechnological applications, such as tissue engineering, drug delivery, wound dressings and medical implants.

Fig. 4 Scanning electron micrographs of **a** raw PALF, **b** steam-exploded PALF and **c** bleached PALF (Cherian et al. 2010)

Fig. 5 ESEM photograph of
the individualized nanofibrils
(Cherian et al. 2010)

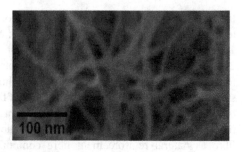

Ledward and Lawrie (1984) isolated different plant enzyme including the pineapple leaf fibre, which have potential use in processing of meat in industrial scale. They showed another direction of preparation of pretentious diet for animals as feed to improve the meat quantity in case of slaughter animals. The processed feed can made through either spinning or thermoplastic extrusion processes. These will improve the blood content of 60–70 % weight of plasma and 30–40 % by weight of suspended erythrocytes red cells. Out of these two components, plasma contains 6 % protein and again out of this protein, about 50 % is serum albumin (by molecular weight). On the line, a team of scientists (Kellems et al. 1979) also worked on utilization of pineapple plant including leaves as fodder and its potentiality as feedstuff for beef cattle. Based on the laboratory experiments and trials, they concluded from this in vitro digestibility, in vivo digestibility, and the feedlot performance trials indicate that pineapple silage and green chop can be utilized effectively in ruminant feeding regimes when properly supplemented with crude protein and minerals. Pineapple plant forage has approximately a 58 % total digestive nutrients (TDN) value, 0.36 % calcium and 0.07 % P on a dry matter (DM) basis. Crude protein (CP) is low (4–7 %) with a digestibility of only approximately 35 %. Non-protein nitrogen accounts for 27–29 % of the total nitrogen. Copper was supplemented to offset the high iron and possibly high molybdenum levels. Pineapple silage and green chop when supplemented with CP and minerals produced a gain of approximately 0.91 kg/day in feeder cattle.

Scientists have made an interesting contribution in application as bioadsorbent (Weng et al. 2009) from pineapple leaf fibre, a new area. The ability of an unconventional bioadsorbent, pineapple leaf powder including fibrous material for the adsorption of methylene blue (MB) from aqueous solution was studied. They found that intra-particle diffusion was involved in the adsorption process and that the kinetic data are fitted well with a pseudo-second-order equation. Fitting parameters revealed that the rate of adsorption increased with decrease in dye concentration and decrease in ionic strength, while the mixing speed did not have a significant effect on adsorption. The adsorption was favourable at higher pH and lower temperature, and the equilibrium data were well fitted by the Langmuir isotherm. The maximum adsorption capacity varied from 4.68×10^{-4} to 9.28×10^{-4} mol/g when pH increases from 3.5 to 9.5. Thermodynamic parameters suggest that the adsorption is a typical physical process, spontaneous, and exothermic in nature. The results revealed that this agricultural waste has potential to be used as an economical adsorbent for the removal of methylene blue from aqueous solution (Weng et al. 2009).

6 Conclusions

It can be concluded that the pineapple leaf fibre which is otherwise known as agro-waste has immense potential to use sustainable in fashion textiles as well as different other industrial applications. A systematic approach needs to be developed for collection of pineapple leaf followed by efficient extraction process. Among the different industrial applications, textile industry including cottage as well as orga- nized sectors, medical textiles, pharmaceutical industry, composite industry, pulp and paper industry is the target industries. As far as commercial aspects are con- cerned, pineapple leaf fibre when converted to fashionable textiles, which is nothing but high-end textiles, has huge demand in the global market. Recent days, the socio-economic sound people globally are giving more preference in the natural fibre-based fashionable textile apparels for which then can afford even much higher price. Lot of promotional systems are being organized globally through fashion show and different advertisement about the sustainable fashion textile products from pineapple leaf fibre. However, apart from fancy and fashion garments, there is huge demand market on curtains, furnishing materials; shopping bag can be developed from this environment-friendly sustainable pineapple leaf fibre. Pineapple leaf fibre-based blended products also have equally demand which can reduce the dependency on the man-made fibres. Blended even the by-product or waste generated during the extraction of pineapple leaf fibre has enough potential in the area of agricultural manure sustainable.

As far as the luxury textiles are concerned, international garment/product brands need to come forward to promote the pineapple leaf fibre-based products so that in the retail market the pineapple leaf fibre-based garments are accessible worldwide to the consumers more easily. This not only popularizes this natural fibre, but also increases the demand of this pineapple leaf fibre, which as a whole the farming as well as product developing community will enrich. Before moving into this, in-depth research is required to improve the different functional and handling properties of such luxury apparel products. Apart from these, series of promotional advertisements are required from different government/public/private sectors of different countries concerning with the plant-based fibres to popularize and provide the importance of pineapple leaf fibre and its luxury textile products to the middle and elite class. These holistic approaches will improve the sustainability of the pineapple leaf fibre-based luxury textiles.

References

Anonymous (2010) Pineapple news. Newsletter of the Pineapple Working Group, International Society of Horticultural Science, vol 17(1), pp 52

Anonymous (2014) Newsletter of the Pineapple Working Group, International Society for Horticultural Science Issue No. 21, July, 2014

Anonymous (2015) http://www.britannica.com/EBchecked/topic/460976/pineapple, Accessed 21 Feb 2015

Arib RMN, Sapuan SM, Ahmad MMHM, Paridah MT, Khairul Zaman HMD (2006) Mechanical properties of pineapple leaf fibre reinforced polypropylene composites. Mater Des 27(5): 391–396

Banik S, Nag D, Debnath S (2011) Utilization of pineapple leaf agro-waste for extraction of fibre and residual biomass for vermicomposting. Indian J Fibre Text Res 36(1):172–177

Chakravarty AC, Sinha MK, Ghosh SK, Das BK (1978) Tensile behaviour of pineapple leaf fibre in wet condition. Indian Text J 85(2):95–99

Chaudhary AK, Singh VK, Tewari M (2012) Utilization of waste agriculture byproduct to enhance the economy of farmers. Indian Res J Ext Edu Special Issue (I):89

Cherian BM, Leão AL, Souza SF de, Thomas S, Pothan LA, Kottaisamy M (2010) Isolation of nanocellulose from pineapple leaf fibres by steam explosion. Carbohydr Polym 81(3):720–725

Devi LU, Bhagawan SS, Thomas S (1997) Mechanical properties of pineapple leaf fiber-reinforced polyester composites. J Appl Polym Sci 64(9):1739–1748

Dey SK, Nag D, Das PK (2009) New dimensions of pineapple leaf fibre—An agrowaste for textile application. In: Shukla Jai Prakash (ed) New technologies for rural development having potential of commercialisation. Allied Publishers Private Limited, Delhi, pp 115–127

Dittrich P, Campbell Wilbur H, Jr Black C C (1973) Phosphoenolpyruvate carboxykinase in plants exhibiting crassulacean Acid Metabolism. Plant Physiol 52(4):357–361

Doraiswami I, Chellamani P (1993) Pineapple leaf fibres. Text Prog 24(1):1–37

Liu E, Guo A, Guo Y, Kong H, Wang Y, Zhang X, He L (2006) Study on enzymatic degumming of pineapple leaf fiber. J Text Res 27(12):41

George J, Bhagawan SS, Prabhakaran N, Sabu T (1995) Short pineapple-leaf-fiber-reinforced low-density polyethylene composites. J Appl Polym Sci 57(7):843–854

Ghosh SK, Dey SK (1988) Can pineapple fibre be used for apparel. J Text Assoc 49(5):167–171

Ghosh SK, Sinha MK (1977) Assessing textile values of pineapple fibre. Indian Text J 88:111–115

Ghosh SK, Sinha MK, Dey SK, Bhadhuri SK (1982) Processing of pineapple leaf fibre (PALF) in cotton machinery. Text Trends 22(10):1–5

Ghosh SK, Dey SK, Ghosh A, Saha SC (1987) Studies on performance of PALF and acrylic blended yarn spun in jute spinning system. Man-Made Text India 30(10):485–488

Hamilton RW, Milgram BL (eds) (2008) Material choices: refashioning bast and leaf fibers in Asia and the Pacific. Fowler Museum at UCLA, Los Angeles, p 187 (ISBN-13 9780974872988)

Kannojiya R, Gaurav Kumar, Ranjan Ravi, Tiyer NK, Pandey KM (2013) Extraction of pineapple fibres for making commercial products. J Environ Res Dev 7(4):1385–1390

Kellems RO, Wayman O, Nguyen AH, Nolan JC, Campbell CM Jr, Carpenter JR, Ho-a EB (1979) Post-harvest pineapple plant forage as a potential feedstuff for beef cattle: evaluated by laboratory analyses, in vitro and in vivo digestibility and feedlot trials. J Anim Sci 48(5): 1040–1048

Ledward DA, Lawrie RA (1984) Recovery and utilisation of by-product proteins of the meat industry. J Chem Technol Biotechnol 34(B):223–228

Li Y, Huang M, Tan H (2004) Chemical modification of pineapple leaf fiber and its application. J South China Univ Trop Agric 10(02):21–24

Liu W, Misra M, Askeland P, Drzal LT, Mohanty AK (2005) Green composites from soy based plastic and pineapple leaf fiber: Fabrication and properties evaluation. Polymer 46(8): 2710–2721

Lopattananon N, Panawarangkul K, Sahakaro K, Ellis B (2006) Performance of pineapple leaf fiber–natural rubber composites: the effect of fiber surface treatments. J Appl Polym Sci 102 (2):1974–1984

Luo S, Netravali AN (1999) Interfacial and mechanical properties of environment-friendly "green" composites made from pineapple fibers and poly (hydroxybutyrate-co-valerate) resin. J Mater Sci 34(15):3709–3719

Mangal R, Saxena NS, Sreekala MS, Thomas S, Singh K (2003) Thermal properties of pineapple leaf fiber reinforced composites. Mater Sci Eng A 339(1–2):281–285

Mishra S, Misra M, Tripathy SS, Nayak SK, Mohanty AK (2001) Potentiality of pineapple leaf fibre as reinforcement in palf-polyester composite: Surface modification and mechanical performance. J Reinf Plast Compos 20(4):321–334

Mishra S, Mohanty AK, Drzal LT, Misra M, Hinrichsen G (2004) A review on pineapple leaf fibers, sisal fibers and their biocomposites. Macromol Mater Eng 289(11):955–974

Mohamed AR, Sapuan SM, Shahjahan M, Khalina A (2009) Characterization of pineapple leaf fibers from selected Malaysian cultivars. J Food Agric Environ 7(1):235–240

Mukherjee PS, Satyanarayana KG (1986) Structure and properties of some vegetable fibres, Part-2: Pineapple fibre. J Mater Sci 21(1):51–56

Nag D, Debnath S (2007) A Pineapple leaf fibre decorticator assembly, Indian Patent Application No. 2334/DEL/2007, 07 Nov 2007

Paul D, Bhattacharyya SK, Banik S, Basu MK, Mukherjee AB (1998) Extraction of pineapple leaf fibre. Appropr Technol 24(4):27

Ranganathan LS, Parthasarathi K (2000) Enhanced phosphatase activity in earthwormcasts in more microbial origin. Curr Sci 79:1158–1159

Ranganathan LS, Vinotha SP (1998) Influence of pressmud on the enzymatic variations in the different reproductive stages of Eudrilus eugeniae. Curr Sci 74:634–635

Saha SC, Das BK, Ray PK, Pandey SN, Goswami K (1990) SEM studies of the surface and fracture morphology of pineapple leaf fibers. Text Res J 60:726–731

Saha SC, Das BK, Ray PK, Pandey SN, Goswami K (1991) Infrared spectra of raw and chemically modified pineapple leaf fiber (*annanus comosus*). J Appl Polym Sci 43(10):1885–1890

Saha SC, Das BK, Ray PK, Pandey SN, Goswami K (1993) Some physical properties of pineapple leaf fiber (PALF) influencing its textile behavior. J Appl Polym Sci 50(3):555–556

Sinha MK (1982) A review of processing technology for the utilization of agro-waste fibres. Agric Wastes 4:461–475

Sinha MK, Ghosh SK (1977) Processing of pineapple leaf fibre in jute machine. Indian Text J 84 (3):105–110

Upadhyay A, Lama JP, Tawata S (2010) Utilization of pineapple waste: a review. J Food Sci Technol Nepal 6:10–18

Uso Y, Dam A (2013) Review on PALF extraction machines for natural fibers. Adv Mater Res 781–784:2699–2703. doi: 10.4028/www.scientifrc.net/A7M8R1.-784.2699

Wang G, Zhang V (2009) The exploitation and the development perspectives of new environmental Foliage fibre. J Sustain Dev 2(2):187–191

Weng CH, Lin YT, Tzeng TW (2009) Removal of methylene blue from aqueous solution by adsorption onto pineapple leaf powder. J Hazard Mater 170(1):417–424

Beyond Appearances: The Hidden Meanings of Sustainable Luxury

Silvia Ranfagni and Simone Guercini

Abstract The notion of sustainability has become increasingly pervasive. It is affecting even luxury sectors where companies attracted by the economic advantages of the green economy develop business models based on sustainable supply chains. Thus, sustainability and luxury, although apparently distant concepts, seem destined to coexist. Their combination appeals to those fashion consumers who, inspired by conscientious consumption, materialize even in luxury goods their sense of environmental and social responsibility. Hence, the importance of investigating what makes a luxury brand sustainable. The aim of this chapter is to identify the dimensions that characterize sustainable luxury and, thus, the sources of the dialogue between luxury and sustainability. The analysis is focalized on four Italian cases (Loro Piana, Gucci, Zegna, and Brunello Cucinelli). Based on the collection of secondary data, it compares company-dominated data (Websites, interviews, etc.) with consumer-dominated data (fashion blogs) by using the text-mining technique. The emerging information produced for each brand has driven us in the interpretation of the individual cases. The main findings show that a sustainable luxury brand is marked by a quality-based ideology, measures itself against a community setting, creates by integrating manual skill with technology, and is part of institutional and political relations. Thus, ideological, community, political, and creative dimensions compose a sustainable luxury. These dimensions suggest that the dialogue between sustainability and luxury unfolds around the features of rigour, perfection, sharing, flexibility, and independence. All these features can be seen as values that a luxury brand should embrace if it decides to turn to sustainability.

Silvia Ranfagni and Simone Guercini share the final responsibility for this chapter. The authors wrote together introduction and methodology, only Silvia Ranfagni wrote the remainder.

S. Ranfagni (✉) · S. Guercini
Department of Economics and Management, University of Florence,
Via della Pandette 9, 50127 Florence, Italy
e-mail: silvia.ranfagni@unifi.it

S. Guercini
e-mail: simone.guercini@unifi.it

© Springer Science+Business Media Singapore 2016   51
M.A. Gardetti and S.S. Muthu (eds.), *Handbook of Sustainable Luxury*
Textiles and Fashion, Environmental Footprints and Eco-design
of Products and Processes, DOI 10.1007/978-981-287-742-0_4

Keywords Sustainable luxury raw materials · Luxury and consumption · Engagement luxury consumer

1 Luxury and Sustainability: An Emerging Path in Fashion Companies

Luxury and sustainability seem to bear divergent, almost irreconcilable meanings. Luxury denotes products and services that generate intensive human involvement, exist in a limited number, and are recognized for their value (Kapferer 1997). Luxury is a hallmark of beauty and an object of desire and, although devoid of utility, produces pleasure, thus becoming also a source of social differentiation: its possession brings esteem, thereby enhancing notoriety, prestige, and status (Pantzalis 1995; Vigneron and Johnson 2004). In short, luxury is inessential, superfluous, exclusive, and ostentatious. In this sense, it is the exact opposite of sustainability (Guercini and Ranfagni 2013). Sustainability is associated with environmental conservation (Berns et al. 2009; Pullman et al. 2009), improvements of social living conditions (Montiel 2008; Closs et al. 2011), and ethical business philosophies (Colbert and Kurucz 2007; Lubin and Esty 2010). Thus, it is an indispensable, collective, and ideological phenomenon. The divergence between sustainability and luxury is not merely terminological. It has also been borne out by specific studies that investigate the perception consumers have of sustainable luxury. Davies et al. (2012), for example, show that consumers consider sustainability and ethics less important in their luxury consumption decision-making process than in their commodity consumption decisions, although luxury products were perceived as more sustainable in comparison with the commodity products. Achabou and Dekhili (2013) go so far as to claim that sustainability embedded in luxury goods can make them less desirable than non-sustainable luxury products. Yet despite this divergence, a growing share of consumers is giving up "conspicuous consumption", which foregrounds individual appearance, and converting to "conscientious consumption", which expresses an evolving environmental sense and social responsibility (Hennigs et al. 2013). As a consequence, international luxury brand companies increasingly invest in sustainability as a means to exploit the consolidated heritage embedded in the manufacturing excellences extending their intrinsic values (Jones et al. 2008; Sharma et al. 2010; Gardetti and Girón 2014). They do this by associating their brands with actions they implement to preserve environmental resources and peculiarities of their original territories and to protect special skills of which their employees are holders. Thus, they express what, according to Kapferer (1997), joins luxury and sustainability, that is durability and uniqueness. Just as luxury goods do not pass in the wake of fashion and animate enduring businesses, so do sustainable behaviours lead to ensuring the continuity of a natural and social environment. Besides, luxury goods derive their value from rare raw materials, also territorially rooted, and craftsmanship competences, just as sustainable actions contribute to not wasting what on the planet is scarce and irreplaceable. Luxury companies seem to make a pact with the environment: they safeguard it and make it the protagonist of their collections. It is a sort of tacit exchange as expression of the alliance existing between luxury and sustainability. What is more, luxury brands that derive from sustainability reciprocal and inte-grative values of their personality attract consumers whose attitudes are consciously ethical (Kapferer and Michaut-Denizeau 2015). These consumers search for luxury brands, which are an expressive synthesis of an individual identity that reflects an empathetic tension and aspiration for a better human environment (Bendell and Kleanthous 2007). They are motivated by a variety of contingent situations. Among them are a rapid and unbalanced economic growth, especially in the emerging countries, and its negative impacts on pollution and on conditions of workforce exploitation, cases of entrepreneurial corruption that damages consumers, and, more in general, a corporate social responsibility, which often appears more declared than real. Those who share a deep and authentic approach to the concept of luxury are critical of companies that ignore the conservation of social and envi-ronmental excellences or that consider it only as a façade of a non-existent brand ethic (Hult 2011). It follows that for sustainable luxury managers, it is essential to understand how to make the relation between sustainability and actual brand values dialectically and as far as possible interpenetrating by creating an aspirational world of brand that is evocative and expressive of concrete and tangible socio-environmental actions. In this chapter, we will analyse the cases of Loro Piana, Zegna, Gucci, and Brunello Cucinelli, which, albeit from different points of departure, have developed sustainable management approaches. On the whole, our aim is to gain an understanding of what constitutes sustainable luxury. Sustainability can be broken down into environmental, organizational, social, and economic dimensions (Guercini and Ranfagni 2013). We seek to understand whether luxury as well consists of the same dimensions or involves still others, and what, aside from rareness and durability, might be other associations that could foster the dialogue between sustainability and luxury. Before proceeding with the analysis of the cases, however, we need to explain how we went about their reconstruction. What follows, then, is a brief methodological note.

2 Methodological Notes

The methodology involves the analysis of business cases constructed from the collection and interpretation of secondary data (Yin 1984; Stewart and Kamins 1993; Guercini 1996). The sources explored are varied and include brand-related press releases found on Websites, non-financial narratives from annual reports, and interviews with house designers published in mainstream media sources. All these company-dominated sources were compared with consumer-dominated sources and, in particular, with brand perceptions emerging from fashion blogs that contain posts written by experts (Guercini 2014; Crawford et al. 2014). The fashion blogs investigated are variegated and selected as the most suitable for the subject of our research. They have been chosen for a variety of reasons. They rank highly according to well-established criteria used to identify successful blogs, including membership, Alexa traffic data, number of indexed pages, and incoming links. Moreover, their posts/comments are also available for approximately three years, which made possible the collection of a sufficient amount of data.[1] For each brand, the keywords used in gathering data are "sustainable", "luxury", and the brand name. All the keywords were searched in combination. We collected data and created overall eight files: (a) four containing for each brand, data that are company dominated and (b) four containing for each brand, data that are consumer dominated. We processed them using the text-mining software T-LAB. This program is widely used by researchers and professionals nationally and internationally (http://tlab.it/it/partners.php). More specifically, we produced a semantic cluster analysis of data (Rastier et al. 2002) to identify themes related to sustainability marked by the combination of common semantic features. We compare for each brand the information resulting from data contained in the files related to company and consumer-dominated data. The combined emersion of information guided us in the interpretation of the individual cases. Finally, these data were also integrated with information on the four companies the authors possess and codified in case studies, as product of five years of research (Ranfagni and Runfola 2012; Ranfagni 2012). The analysis delineated the dimensions of sustainable luxury and their contents, which we will present in the discussion section on the sustainable luxury dimensions.

3 Cases of Italian Luxury Brands

In this section, we will analyse the four cases of Italian companies. Brunello Cucinelli is the youngest of the cases investigated. It has developed a sustainable approach, which is filtered by a humanistic vision of business. This makes it an interesting case when compared with the other three. All the cases even if they do not necessarily belong to international sustainability classifications are recognized by consumers participating in fashion blogs as sustainable brands because of their actions in defence of the environment and social resources.

---

[1]Here, we list some of the main fashion blogs investigated. For Zegna: cashmereworldfair.com, fashionwelike.com, bluandberry.com, evermanifesto.com, vegasmagazine.com, and fashionnvest. com; for Gucci: earthtimes.org, eco-age.com, ecoblog.it, fashion.telegraph.co.uk, greenbiz.it, greennew.it, ideegreen.it, luxurysociety.com, and urbantimes.co/; for Loro Piana: ammando.com, ecouterre.com, harryrosen.com, knittingindustry.com, and mcadsustainabledesign.com; and for Brunello Cucinelli: campdenfb.com, enablingideas.com, fashionwelike.com, fismo.it, and treedon. net.

3.1 Loro Piana: Naturalness Generating Sustainable Luxury

Loro Piana is a company that has achieved a leading position in the production of cashmere, vicuña, and extra-fine wools. Behind the company is a family business, which began its activity as wool merchants and founded in the second half of the nineteenth century a woollen mill at Quarone in Valsesia (region of Piedmont). There is situated the company's corporate headquarters, completed in 1924 by the engineer Pietro Loro Piana. In the period following the Second World War, Pietro's grandson took over the family business and turned it into a well-known fashion brand. Today, the company has 2533 employees and two divisions. The first is the wool factory, which manufactures high-quality textiles using noble fibres such as cashmere, vicuña, and extra-fine wools. The second is the luxury goods division, which produces a complete line of clothing products and accessories for men, women, and children using the company's own fabrics. Loro Piana is a vertically integrated company and, as such, has complete control over the whole supply chain; thus, the luxury it produces is built on a continuous and monitored quality. The search for quality has animated every generation of the Loro Piana family; each has managed the company inspired by the values of excellence and the respect for the business history, especially in terms of manufacturing traditions. The quality the company pursues involves a meticulous attention to fine raw materials. The main sourcing areas are Australia, New Zealand, Mongolia, and China. While in Australia and New Zealand, it seeks the highest quality fleeces from merino sheep, in Mongolia and China, it searches for rare fleeces from local goats. The natural environments where Loro Piana finds its finest raw material are unspoiled worlds that supply inimitable resources. The Loro Piana family, in fact, invests economic and human resources in order to identify and, then, to preserve them. This is because they are convinced that it is precisely thanks to them that it is possible to combine quality and timeless elegance and to produce fabrics and garments destined to last more than a lifetime. A quality-based philosophy seems to inform the search for yet unexplored natural resources. Their preservation at the base of Loro Piana's sustainable orientation is thus coherent with an existing company way of being. All this explains the recent environmental projects that the company has decided to undertake. One of these concerns the opening of a subsidiary in Ulan Bator (Mongolia) to work together with nomadic tribesmen in the breeding of goats whose fleece is used in the production of cashmere clothing collections. In particular, the aim of this green field investment is to transfer to the local community of breeders all the techniques of animal husbandry in order to preserve a local ecosystem and, at the same time, to pursue the high quality of the emerging raw material. The survival of this latter depends on the conservation of the underlying rare resources rooted in naturalistic environments. The products Loro Piana realizes by using Mongolia fleece are a source of pride; particularly cherished, and appreciated by consumers, are the products labelled as Loro Piana Baby Cashmere. Their fibre comes from Hyrcus goat kids that are between three and twelve months old.

Marked by an unmatchable softness, its derivatives include limited editions of long overcoats (average price $50,000 per item) along with baby clothes such as body suits, caps, and socks. Overall, Loro Piana cashmere is a real luxury; its value in terms of price to pay makes it accessible only to market niches. Since it is made of raw materials that are deeply natural, its products seem to appeal mainly to consumers who understand and recognize their untouched origin. They are in markets such as Europe, the USA, and Japan. However, also consumers who belong to emerging countries are revealing themselves an interesting segment. Some of them, in fact, are beginning to express a higher sensitivity to quality perceived as such on the basis of the manufacturer's country of origin (Loro Piana collection are 100 % made in Italy), independently of the underlying brand awareness. The perceived value of Loro Piana collections depends on raw materials, manufacturing place, and also on production processes. Loro Piana has, in fact, invested in the most up-to-date technologies, which carefully and thoroughly process fibres to make them to achieve the highest level of refinement and quality. The technologies do not alter raw materials but exploit to the greatest extent their innate potential. However, the high quality in the finished products is due to a combination of technology with human competences and, in particular, with superior craftsmanship, necessary to make handmade adjustments.

The search for new productive experiences involving natural communities has led Loro Piano to develop agreements with foreign governments. Thus, sustainable choices the company makes coherently to its ethics of quality pass through political relations. A case in point is the ongoing collaboration, since 1994, with the Peruvian government to save the vicuña, the animals that produce the most precious and finest natural fibre (12–13 microns against the 15 microns of cashmere). Since the vicuña was at risk of extinction, Loro Piana was granted permission to acquire 2000 ha of local land to be converted into a private reserve, where the animals could live protected from poachers. In exchange for this intervention, Loro Piana has enjoyed preferential access to the reserve and, in particular, has obtained exclusive rights to the purchase of 85 % of the fibre produced. Now, Loro Piana preserves unique resources including them in collections, which thus emanate from a process of company integration with foreign communities of producers. These communities, although faraway and culturally distant, share with Loro Piana an important objective: the safeguarding of their environment together with their local traditions. The extremely soft golden fur of the vicuña has, in fact, a centuries-old history. "The Princess of the Andes", as it is called, was once reserved exclusively for Incan Emperors; its loss would be considered the death of an ancient Peruvian symbol. Overall, the raw fibre obtained from vicuña amounts to less than 5000 kg, as compared with the 10 million kg of cashmere and 500 million kg of wool. The dedication of Loro Piana to natural resources is witnessed also by an experiment carried out on lotus flower fibres, a natural raw material produced on the lakes of Myanmar. These fibres, once extracted from the plant, have traditionally been spun by hand and woven to produce a fabric, akin to raw silk, which in the past marked local textile traditions. Loro Piana decided to use this fibre and the resulting fabric to produce a special jacket.

Driven by the mission to sell excellence made from the best sources of raw material, the company increased its sales from Euro 242 million in 2000 to Euro 630 million in 2012. It is globally distributed in 22 countries and operates 135 retail stores, most of which directly owned. A quality-oriented approach, integration of local communities, scouting activities, and experimentations, has favoured in Loro Piana the union between luxury and sustainability. The driver behind all of this is the passion of a family, which, at least until some years ago, has believed in its sustainable projects by exploiting them to enhance the manufacturing excellence underlying the company identity. In addition to passion, the success of sustainable luxury in Loro Piana is also due to sophisticated marketing skills, capable of transforming the exclusive raw material into a visible source of brand differentiation. Both Baby Cashmere and Vicuña are now premium Loro Piana labels that consumers associate with products of outstanding and unique quality. However, something has marked the recent history of Loro Piana. If on the one hand, it is characterized by continuity in sustainable investments, and on the other hand, it is tinged with changes in the company-ownership. In 2013, Loro Piana acquired for 1.6 million dollars the majority of a company that has the right to shear about 6000 vicuña in Argentina. In the same year, it also sold its 80 % to the French Group LVMH and remains affected by the death of Sergio Loro Piana, who, together with Pier Luigi Loro Piana, composed the company's top management. In light of these events, Loro Piana will be an interesting case to observe in the near future. We are dealing with a family business, which, after having fostered the cult of quality through sustainable experiences, finds itself, for contingent reasons, passing the baton to a multinational of luxury. It is a delicate and insidious choice that among consumers raises questions as to the possibility of preserving sustainable values embedded in a quality production without becoming mere subordinates of economic and financial interests. What is perceived to be at risk is the human integrity that has fuelled the true brand ethics, and with this, the passion of those who created it and made it tangible in the eyes of consumers. Thus, the question that arises is whether sustainable experiences such as those of Loro Piana, once they are incorporated in a multinational company, can really survive over time and not just become a memorable case destined simply to be narrated.

3.2 Gucci: Formalism as the Basis of a Sustainable Corporate Approach

Gucci is a luxury fashion company, which has recently made sustainability as an integrating value of its brand. The development of a sustainable business approach in a multinational company like Gucci, that is recognized in international markets, structurally consolidated and integrated in a predefined network of suppliers, was far from easy. It required formalized relational processes capable of activating in both internal and external company actors a managerial sustainable orientation. The engine of these processes has been the profound sensitivity Gucci has always held for the manufacturing excellence together with an increasing awareness that luxury fashion consumers are reconsidering their relations with brands. Since they are inspired by more authentic and, thus, realistic consumption ideologies, they tend to judge luxury brands combining their intrinsic quality with their involvement in actions related to social and, more in general, environmental issues. In Gucci, sustainable actions were preceded by formal acts endorsing sustainable corporate ethics aimed at searching for the highest level synthesis of brand quality through natural and human resources protection and preservation. Gucci's conversion to sustainability can be traced back to 2004, when the company initiated a process of certification in the corporate social responsibility (SA 8000). This certification regards leather goods, jewellery, footwear and clothing supply chains, and stands for values such as business ethics, respect for human rights, and workers' health and safety. The process, certified in 2007, was only the first stage in Gucci's quest for sustainability. Another important step was the special formal agreement signed in 2013 with the Ministry of the Environment and Protection of Land and Sea to set off a process of reducing CO_2 emissions along the manufacturing supply chain. This agreement concerns the assessment of the environmental impact in terms of "eco-costs" to verify whether production is certifiable as sustainable according to specific international standards. The quality-based brand value pursued through a sustainable attitude leads Gucci to make the underlying socio-environmental sentiment a unifying feature of the company: once shared, it can bestow a deeper meaning on the achievements of both employees and economic actors that have built consolidated business relations with the company. As bearers of distinctive technical skills, employees and manufacturing partners are seen by Gucci as a community of workers fostering a real heritage to be preserved over time. The need not to lose it motivated the company to sign, in 2009, an agreement with business associations (Association of Industries and Local Craftsmen Confederation) of Florence (Tuscany) to ensure a formalized exclusive relation with the company partners. Many of them are located near Florence where Gucci has also its head-quarters. Thus, Gucci has created the conditions to count on the local partners' expertise, which lies not only in their craft skills, but also in the ability to convert them into sustainable production practices that are internationally recognized. By means of flexible and specialized human resources, it can expand its sustainable experiences to foster processes of product innovation based on the use of new manufacturing technologies. Environmental projects have served Gucci to be creatively more inspired and filter brand values making them more credible and, in certain sense, more pure. Instances attracting market attention are the new packages (100 % recyclable paper) designed in 2010 to limit the use of material and produced exclusively according to FSC (Forest Stewardship Council) guidelines, and more recently, the completely (100 %) traceable handbag collection realized in partnership with the well-known eco-fashion advocates Livia Firth. Sustainable innovations also involve the manufacture of glasses. In 2011, in fact, Gucci realized in partnership with Safino eyewear models using an innovative acetate, which, compared to the traditional ones used for optical frames, contains a much higher percentage of material of natural origin. Consumers seem to appreciate especially the sunglasses Gucci has made from liquid wood, a biodegradable, eco-friendly material alternative to the plastic normally used in the production of eyewear. Liquid wood is composed of bio-based materials: wood fibre from sustainably managed forests and lining from the paper manufacturing process and natural wax. Innovation, driven by the quest for sustainability, also led the company, in 2012, to make a special edition of eco-friendly women's ballerina flats and men's sneakers. The former are made of bio-plastic, while the latter combine bio-rubber soles, biologically certified laces, and a top in genuine vegetable-tanned black calfskin. Gucci's sustainable orientation finds strategic coherence in the strategies implemented by its membership group that is the Kering Group. This latter in fact pursues corporate social responsibility in the luxury sector. In particular, it intends to eliminate hazardous chemicals from its production by 2020. Its action plan focuses on the reduction of carbon dioxide emissions, waste and water, sourcing of raw materials, hazardous chemicals, and paper and packaging. The formalized and collective-based sustainability orientation Gucci has developed seems to have contributed to increasing its economic results. Gucci is now a multinational company with revenue of over 3.1 billion Euros, more than 8000 direct employees and over 370 directly operated stores around the world. One of its most widely recognized expression of social and environmental initiative concerns the creation of the world's first bag certification as zero deforestation from Amazon leather. It has involved natural communities of local workers as guardians of rare resources. Its production took two years. This is the time to bring the bag from concept to creation. The Gucci Environmental Responsibility Manager, Rossella Ravagli, and her specialized team lived on a ranch in Brazil to experiment with local communities how they could raise cattle without chopping down a single tree. This experimentation is seen as a way to reduce an environmental issue whose impact is both local and global. Brazil has the largest commercial herd of cattle in the world, and grazing land for cattle in the Amazon is now the driving factor behind 75 % of tropical deforestation. It is not just a question of the damage and decimation of plant life; deforestation is also one of the major causes of climate change, as it contributes to approximately 20 % of the world's greenhouse gas emissions. The leather Gucci uses to produce bags is sourced from Fazendas Sao Marcelo, Ltd, a group of four ranches located in the Mato Grosso state in Western Brazil. The ranches cover a total area of 79,000 acres (32,000 ha) including a reserve of 32,000 acres (13,000 ha) in the Amazon. All these ranches have earned the Rainforest Alliance Certified seal of approval. This means that they curb deforestation, protect wildlife habitats, provide ethical treatment to livestock, and promote the rights and well-being of ranch workers. This shows that environmental sustainability and social sustainability converge in the labour of a natural community: its inhabitants contribute to the protection of the environment where they live, at the same time improving their quality of life. What Gucci's former creative director, Frida Giannini, has to say in this regard is emblematic: "At Gucci we would like not just to be synonymous with made-in, but also made-with-integrity. In this way, we try to meet the green criteria without compromising Gucci's reputation for luxury" (http://fashion.telegraph.co.uk/article/TMG9902615/Eco-fashion-Why-green-is-the-new-black-for-Gucci.html—accessed, 12 February 2015). The Amazon community is an example. Here, Gucci tries to create community integrity through environmental and social conservation. Thus, Gucci brand could enrich its expression of exclusivity through sustainable experiences based on local integration. By doing this, the company contributes also to preserving all the traditions and cultures related to natural communities. In sum, Gucci stands as a company that has been able to formalize through the definition of common rules a sustainable orientation, contaminating both those who work in and those who work with the company. It succeeded in this company despite the fact that it is a multinational and, as such, subject to continual financial pressures, which, at times, inhibit effective, ethically responsible behaviours, and, as a result, the ability to listen to emerging market segments as pioneers of new needs. Experts recently ask, however, whether the increasingly aggressive competition and the company's recent organizational overhaul (with the exit of the creative director) will affect the sustainable path that Gucci has embarked upon. A formalized sustainability can perhaps be interpreted as a first step in a sustainable process that must be continually fuelled and renewed, while maintaining a balance between differentiating sustainable values and the continuum of brand values. This is the challenge that today characterizes many sustainable luxury multinationals. Sustainable luxury is increasingly becoming a source of competition between companies, so it may well take on new features in the near future. Regardless of the changes, it is desirable that it retains over time a core of authenticity without turning into mere trimmings of a brand appearance. Otherwise, it could lose its ethical purpose and, thus, its underlying *raison d'être*.

3.3 Zegna Group: Sustainability as a Connecting Link Between Businesses

The Zegna Group is a family company, established 1910 in Trivero, in the Biella Alps (region of Piedmont). Its founder's objective was to create the world's finest innovative fabric, sourcing the noblest and the rarest fibres directly from their markets of origin. He seems to have been successful in this plan. Zegna fabrics are, in fact, recognized as among Italy's most prestigious. The intent to employ these materials in independently designed collections has led the company, since the 1980s, to pursue a strategy of vertical integration. Managed by the fourth family generation, Zegna is today a global luxury brand, which includes fabric, clothing, and accessories production. The internationalization process it has triggered involves both production and distribution. Zegna has opened production subsidiaries in countries such as Spain and Mexico, and following the inauguration of the first flagship stores in Paris and Milan, it has invested in retailing and, well ahead of its competitors, has reached out to the emerging markets. In particular, it was the first international luxury brand to open a store in 1991 in Beijing and to make China its principle market with over 70 directly owned stores. Overall, Zegna is present in over 80 countries with 555 stores, 311 of which managed directly. The need to acquire new skills in business areas, which it does not dominate, like leather goods, led the Group to purchase, in 2002, the brand Longhi, a clothing manufacturer of luxury leather, and to forge a joint venture with Salvatore Ferragamo (ZeFer), one of the international luxury brands best known for the production of footwear and bags. In its growth, the company has acquired new skills, but at the same time, it has undertaken social actions to retain and, thus, preserve the internally existing ones. A demonstration of this is the decision to invest in building collective complexes near the company headquarters where workers can decide to live by exploiting the workplace proximity. Besides, thanks to Zegna's economic efforts, since the early 1930s, employees, but also Trivero citizens, can make use of a library, a gym, a theatre, a public swimming pool, a medical centre, and a kindergarten. All these initiatives produce a sense of community, both within the company (among employees) and in the social fabric (among local citizens). Thus, Zegna strives to enter the collective imagination of the local communities it contributes to generate by building up with them relations that are trust-based. In particular, these relations once developed with employees can act as a kind of intangible barrier against the drain of internal competences, which are amalgamated and refined through continuous processes of productive experimentation. There follows a sense of social sustainability underpinned by taking positions on social issues involving citizens and workers. These latter are seen by Zegna Group as an irreplaceable source of differentiation; the more refined their skills, the rarer they become, and the greater the need to preserve them. The perceived value of the brand, in fact, seems to depend not only on the nobility of the raw material, but also on the expertise of the underlying human resources. In Zegna, social sustainability is combined with a marked environmental sustainability. In fact, as just described, the company acts to preserve the environment by making it a shared good that positively affects the collective quality of life. However, starting from the company's foundation, the environmental commitment has been profuse. In love with the beauty of nature and deeply attached to the land, its founder gave rise to an impressive project of environmental enhancement, creating a genuine natural park known as "Oasi Zegna". In this regard, Angelo Zegna, current Honorary Chairman of the Group, likes to point out that "my father was an environmentalist long before the term came into existence" (http://www.zegnagroup.it/pdf/en/history_and_development.pdf—accessed 12 February 2015). The Oasi Zegna covers approximately 100 km^2 between Trivero and Valle Cervo, in the Biella Alps, and is the first Italian example of "environmental heritage". For every new born child of the Group employees, a new tree will be planted in the Oasi Zegna. This oasis is a natural site and a protected park that offers visitors an educational and emotional experience. It stands as a real, open-air "laboratory" for the new generations and is seen as an ideal place for families, children, and outdoor sports enthusiasts. Through the Oasi project, Zegna is trying to transform the protection of environmental uniqueness into a source of economic aggregation. In fact, it has established a Tourism Consortium, which includes selected local actors, such as hoteliers and sports and cultural operators, in an effort to develop a structured tourism and recreation supply. Not merely a nature park to visit, but a genuine cultural tourist destination capable of accommodating a national and international audience who are sensitive to the values of environmental conservation and environment-friendly sporting activities. Thus, the distinctiveness Zegna seeks lies not only in human skills and in the raw material marking its fashion-based supply, but also in the combined tourist and environmental contents, which are specific to its leisure-based supply. Ultimately, it protects environmental singularities and turns them into a source of new business, that is sustainable tourism. All the activities Zegna has included in the Oasi's valorization plan, aimed at honing the skills involved and coordinating the different environmental attractions, are decided in partnership with local institutions and associations. Thus, as promoter of local territorial development, Zegna can manage the Oasi boasting also an institutional legitimation.

The sustainable behaviours involving Zegna's core business translate into products resulting from manufacturing processes that respect the European environmental standards as well as from resources that, as they are in risk of extinction, are socially protected. The production of knitwear yarn, for example, is assigned to a specialized staff that carries out environmental tests to optimize energy and water consumption. Further, specific collections are created using the noble vicuña fibre, to which Zegna has access as member of the International Vicuña Consortium. This role is due to collaborative relations Zegna has developed with the Peruvian government to implement a project designed to create a water system capable of supplying the local communities of vicuña breeders. To the extent to which these local breeders protect their livestock from poachers, they are entitled to the proceeds from the sale of the shearing going to the companies, like Zegna, that contribute to the safeguarding of their environment. In terms of sustainable behaviours, Zegna also realizes products made from materials, which, if scattered in the environment, can become a source of pollution. Their value lies not in a preserved uniqueness but in recycling. The company, under the Zegna Sport label, has recently realized a short city jacket composed entirely of reprocessed plastic materials. Water bottles and soft drink bottles not used after their normal life are exploited in manufacturing processes. Zegna Sport has also produced a solar-powered jacket whose outer fabric, breathable membrane, seam tapping, lining, and padding are all derived from recycled plastic. Zegna is a case of a family company, which has developed sustainable behaviours making the protected environment also as a source of new business. We are dealing with a model of combined environmental business: sustainability brings diverse businesses together and constitutes a connecting link. The ensuing relationship involves both giving and receiving: you protect the environment so that you can exploit economically what it yields. This also expands the confines of brand. Its association with more than one sustainable business increases its notoriety and underpins its ethical status. This behaviour is strategically interesting, but can conceal a risk: making sustainable behaviour mere instruments rather than goals of a management philosophy. It is precisely the relationship between means and ends that luxury companies should call into question.

3.4 Brunello Cucinelli: Human Dignity as Engine of Sustainable Behaviours

Brunello Cucinelli is a luxury company located in the centre of the country (Umbria) and specialized entirely in cashmere production. It was founded recently, in 1979, on a precise business concept: to exploit local manufacturing traditions and skills to work the noble cashmere from India, Mongolia, and China and to produce colourful, oversized pullovers for women. This idea distinguished Brunello Cucinelli from the traditional producers of cashmere, who focused mainly on the creation of men's collections in conventional colours. In addition to this, what especially marks the company is the development of a humanistic managerial approach. It adopts a way of organizing the social fabric, within the company, based on mutual respect. Its roots can be found in a unique and, in some ways, auto-biographical managerial style. The owner, in fact, declared "I have always culti-vated a dream: to make man's labour more human" (http://www.fashionwelike. com/conversations/brunello-cucinelli-sustainable-luxury—accessed 11 February 2015). Conditioned by the experience of his father, who went from being a happy, cheerful farmer to becoming a depressed, anonymous worker in a cement factory, he decided to develop a luxury fashion business that endorses human ethics and moral values. Selling luxury in Brunello Cucinelli was to be like selling respect for human dignity. This makes the company a distinctive social sustainability case if compared with the other explored luxury brands. In fact, its sustainability, rather than social, can be more precisely defined as *human*; this is because it aims to improve and maintain over time the human self-respect of a local community which embraces not only employees, but also all those economic actors (first and foremost suppliers and consumers) that interact with the company. One of the consequences was the decision to grant employees a monthly salary 20 % higher than the rate set out in the national labour contracts. There is no doubt that this choice, when it was taken, was countercurrent: Brunello Cucinelli preferred to satisfy economically his employees, and extol their skills, rather than outsource production and take advantage of foreign labour, less expensive but not specialized. However, the motivation of this choice is also due to the increasing perceived value of the brand and thus to the related *premium price* recognized by consumers. In addition to compensation policies, the human-based sustainability of Brunello Cucinelli has oriented other collective decisions. First, the creation of a protected labour envi-ronment is far from industrialized society. The company is located, in fact, in the medieval village of Solomeo in the Umbrian countryside (near Perugia), which includes a fourteenth-century castle, medieval houses, and farmhouses rich in spiritualism and mystique, where employees can work and live. Then, the building in the same village of recreational spaces such as a restaurant, open for employees every day, and of a Theatre and an Arts Forum, whose events and performances are available also for local residents. Finally, the definition of management rules, rigorous yet tolerant, that foster homogenous behaviours within the company. For example, the employees are not required to punch a card when they enter and are free from constraints of a hierarchical nature, since there is no clear distinction between management and non-management areas.

It follows that in Brunello Cucinelli, a human-based sustainability can pass through an environmental sustainability. In other words, it results in actions benefitting employees that are direct (e.g. higher salaries) and indirect. The indirect ones translate into a recovery and a conservation of natural environments. More specifically, a climate of mutual respect is realized preserving the natural environment and then making it a shared resource. The effects are positive. The company life that takes place in a collective natural space if managed through internal rules can contribute to generating a motivated and cohesive employee community. The feeling of unity is also increased by a convivial entrepreneurial behaviour. The owner, in fact, likes to state: "I often meet with the employees to talk about how the company is doing, the profits, my ideas, the stores we are opening. I share everything with them" (http://www.enablingideas.com/how-did-they-do-it/2012/04/brunello-cucinelli-its-not-about-me-as-a-brand-its-about-humanity/—accessed 11 February 2015). The sense of sharing and of listening Brunello Cucinelli develops, and finds its root in his life experience and in his passion for philosophy. His inspiring and favourite philosophers are Socrates, Plato, Seneca, Saint Augustine, Saint Francis, and Marcus Aurelius, the second-century Roman emperor. They help him to think in depth without being overwhelmed by instinctive actions that can turn out to be an end in themselves. They inspire him in developing a human-based sustainability that seems to find its complementation in environmental sustainability. Their combination allows the company to foster actions that positively affect a broader context that is society. Thus, the collective involvement it produces extends beyond an internal community and underlies a company search for a social legitimation of its actions. Proof of this is the so-called projects for beauty the company is involved in. It has supported and financed parks in Salomeo, the Laic Oratory Park, and the Agricultural Park. The first is equipped with a stadium without barriers where young people can train and play in natural surroundings. The second consists of 70 ha dedicated to vegetable gardens, orchards, olive groves, and corn, which will supply the company canteen and the local population. Thus, Brunello Cucinelli has gained an inimitable strategic advantage based on human management as driver of an environmentally and socially oriented business model. Similar to what has occurred for other fashion companies, its sustainable behaviours seek social, but also political legitimacy through the development of relations with government organs. Recently, for example, Brunello Cucinelli has expressed its willingness to assess carefully the CO_2 emissions in its supply chain with the aim of bringing them down in line with international norms and standards. The company has formalized this commitment in an agreement signed with the Italian Ministry of Agriculture. The results of this behaviour are positive. From a single store in Porto

Cervo, a resort town in Sardinia (an autonomous region of Italy), the company has grown to 60 stores in five continents and 1000 points of sale, if non-company operated stores are included. The company achieved an outstanding revenue growth of sales, which in 2014 reached the level of 355.8 million Euro. Currently, Cucinelli sells 70 % of its coloured sweaters abroad, in 59 single-brand boutiques and 1000 multi-brand shops. The principle markets are North America and Europe. The company's products are made in Umbria; the workforce consists of 700 employees and a network of micro-companies employing about 2300 individuals. The case of Brunello Cucinelli stands out as a particular approach to sustainable luxury that consumers recognize although it has not yet led to the company's inclusion in the main relevant international sustainability rankings. Unlike the other brands explored in this chapter, Brunello Cucinelli sustainable approach is the result of a shorter company history, begun, however, with a specific aim: that of preserving a social community based on human dignity. Now, despite this differentiating approach, we question—together with some consumers—the possibility of its conservation over time. The question is meaningful for a variety of reasons. First, the humanistic approach once it has been filtered by an environmental sustainability has given rise to actions ("projects for beauty", Ministry agreements) capable of generating a brand echo both in social and political contexts. But it has also recently favoured an effective approach of the owner to institutional circles as supporter of political actors. This choice could appear to reflect an opportunistic behaviour, which, if real, would undermine the credibility of the human approach as it has been conceptualized. Moreover, Brunello Cucinelli's is an owner-based approach. Thus, as such, it could not be simple to transfer and maintain it independently of its founder. Indeed, it would require exploiting the managerial heritage and carrying it over to new entrepreneurial generations through a process of *entrepreneurial sustainability*. This process is what is lacking today in many medium-sized Italian companies that are struggling to preserve and foster the core of a business philosophy generated by the visionary perspective of their founders.

4 The Emerging Structure of Sustainable Luxury

Social sustainability and environmental sustainability are intertwined in the cases investigated thus far. It may occur that social sustainability passes through the environmental: social rights can be preserved making the environment a collective good. In Brunello Cucinelli, the human sustainability approach pursues the protection of personal dignity, giving rise to a more circumscribed interpretation of social sustainability, which also in this case is pursued in part through the sharing of reclaimed and subsequently protected natural environments.

At this point, we can attempt to abstract from the cases the dimensions of sustainable luxury. One of these is the ideological one (Fig. 1). First of all, an indispensable prerequisite for making luxury sustainable is a deep-rooted *ideology of manufacturing excellence*. Its promoter is usually the company founder. This

Ideological dimension	LP	G	Z	BC
Manufacturing excellence-based companies	x	x	x	x
Being quality lies in the intrinsic values of its components	x	x	x	x
Search for uniqueness in raw material and competences	x	x	x	x
Sustainability as a process of identity rediscovery		x		

Creative dimension	LP	G	Z	BC
Conservation of handicrafts as collective heritage	x	x	x	x
Integration of manual skills with technology	x	x		
Adaptation of standardized production emerging from advanced technologies	x	x		
Modernity and tradition values coexist in brand identity			x	

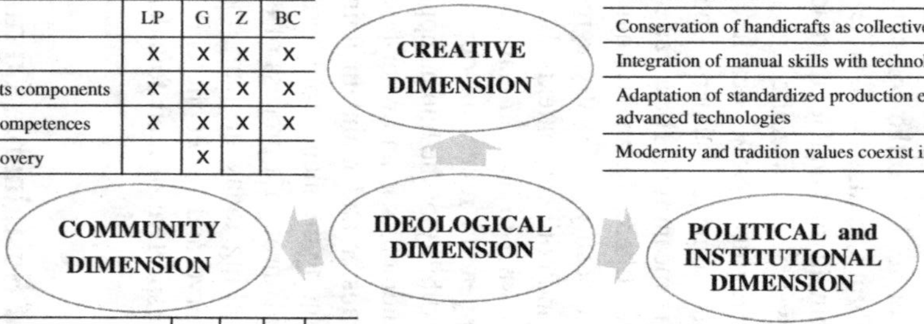

Community dimension	LP	G	Z	BC
Connective environment		x	x	x
Natural communities of producers		x	x	
Listening, mutual understanding, desire to build	x		x	
Artificial communities of producers	x	x	x	x
Consumer communities (sustainable luxury as selective luxury)	x	x	x	x

Political and institutional dimension	LP	G	Z	BC
Sustainable luxury and "polis"	x	x	x	x
Diplomatic soul of sustainable luxury	x	x	x	x
Political legitimacy a priori (activate sustainable processes) or a posteriori (formalize sustainable processes)	x	x	x	x
Social institutionalization		x	x	

CREATIVE DIMENSION

IDEOLOGICAL DIMENSION

COMMUNITY DIMENSION

POLITICAL and INSTITUTIONAL DIMENSION

Legenda: LP = Loro Piana, G = Gucci, Z = Zegna, BC = Brunello Cucinelli

Fig. 1 The dimensions of sustainable luxury from cases analysis (*Source* Our results from secondary data collection)

ideology underlies sustainable luxury and is common to Loro Piana, Gucci, Zegna, and Brunello Cucinelli. It fuels the quest for *uniqueness in using raw materials and competences* that are scarce and that become accessible to those who, ultimately, are able to preserve them. In particular, the competences assume a sacred value for the company; in Brunello Cucinelli, for example, their safeguarding may imply managerial approaches geared to protecting even the innate right they conceal, that is human dignity. In all the cases explored, the ontology of quality and, thus, being quality is fostered by the *intrinsic value* of what contributes to its composition: the searched for rarity incorporates distinctive qualitative contents that make it super-lative and almost universal. The search for quality in rarity activates a natural process of preserving social and environmental resources. Thus, sustainability takes its place in a company ideology, becomes an expression of it, and promotes a tangible-adaptive metamorphosis: in sustainable products, excellence materializes in renewed products that are the result of a productive experimentation. In Gucci, being sustainable becomes a condition that triggers processes of *identity redis-covery* by intensifying or reviving the desire to innovate and to radicalize the sources of a pursued differentiation. The resulting renewal affects not only those who pursue it, but also those with whom they interact. This is because excellence sought through sustainability lays down requirements and productive rules com-pany employees have to respect together with the actors of the supply chain involved. In this way, a *sustainable culture*, which starts out *individual*, becomes *collective*. We are witnessing an ongoing process of contamination whose engine is a shared respect for socio-environmental issues and an underlying sense of rigour and perfection. What is produced can gain value and thereby a sense that exists regardless of the brand awareness; however, to the extent that the "meaning-value" of the product combines with that of the brand, sustainable production and the relative brand develop an interpenetrating dialogue in which each strengthens the other. All this is plausible in the case of sustainable behaviours that are genuine. If they are genuine, they do not conceal a fictitious identity but are rather a true expression of a coherent and rigorous brand ethic rooted in an authentic business identity. Sustainability can thus be seen in this case as a stage in the pilgrimage the company undertakes towards the discovery and the emergence of its true self.

The ideological dimension precedes and is a prerequisite of the other dimen-sions. One of these is the *community dimension* (Fig. 1). Sustainable luxury appears as a *connective environment*, where needs, values, and cultures are collectively shared. In fact, Gucci, but especially Loro Piana and Zegna, interacts with com-munities, varied in nature but all custodians of unique resources. These commu-nities need to be discovered, reassembled, and managed. Their members are *producers* of natural resources located in distant lands, pristine, detached from the hustle and bustle of the modern world. They are often breeders attached to their lands and proud of them. The role they assume is that of irreplaceable partners for the companies investigated. Loro Piana, Gucci, and Zegna transfer them the tra-ditional sustainable breeding techniques or contribute to the preservation of local ecosystems and the underlying traditions, obtaining in return prestigious primary resources. This interchange requires *listening, mutual understanding,* and the *desire to build* something *together,* despite geographic and cultural distances. The sought-for uniqueness is also a cultural discovery. In fact, it is created in unimaginable places, where feeling counts more than action and where history and traditional ritual customs animate collective realities by transforming them into nuclei that are stable and resistant to the forces of modernity. These nuclei, once internally organized, interface with *artificial communities,* created according to a voluntaristic logic based on predefined requisites. These communities join Loro Piana, Gucci, Zegna, and Brunello Cucinelli and consist of internal employees and of specialized suppliers–holders of special competences. Like the natural ones, they have something in common: they share the same sustainable project. The production of sustainable luxury takes shape, then, within community environments and not outside of them. The same is the characteristic of sustainable *consumption* as it emerges from the four cases explored. It, too, seems to manifest itself within predefined boundaries. Its actuators are not dispersed, but united by the same creed: contributing to the preservation of the earth and its peoples. In this sense, they form *communities,* which are visible and therefore identifiable in the heterogeneity of markets. Thus, the identity of a sustainable brand may strengthen its perception in the consumer's mind, if it is able to transfer that sense of collectivism underlying its ethical values. The sustainable brand thrives on deep, shared convictions that act as factors contributing to social aggregation. The luxury that characterizes it, precisely because it is authentic, seems to address those who recognize and understand it. It is, therefore, a selective luxury.

Sustainable luxury requires *manual skills* combined with *technology.* It is just this combination that fosters the creative dimension of sustainable luxury (Fig. 1). All the four cases explored show that sustainable luxury does not take shape without the appropriate craftsmanship. This latter is an inalienable resource on which the value added of the final product depends. The *ductility,* the plastic capacity, and the experience incorporated by craftsmen promote processes of experimentation and, with these, the *adaptation of raw materials* that are natural, never transformed, and processed, to the company's creative projects. Those who possess these skills contribute to giving shape to what the environment offers, breathing life into innovative products in which naturalness, expertise, and style interact in harmony and reinforce each other. However, *manual skills* alone cannot guarantee full meaning. As it derives from Loro Piana and Gucci cases, their achievements are in fact ensured by a continuous dialogue with what appears to be their opposite, that is technology. Yet in sustainable luxury, even opposites seem to find some form of integration. An advanced *technology* allows artisans to express the full potential of their manual flexibility. It materializes all the distinctive qualities of the processed raw materials, transforming them into standardized semi-finished products that are then easily malleable in different models of finished products. Sustainable productions are ultimately individual works of art, the result of handicrafts that the company tries jealously to maintain over time. It preserves them not alone, but together with what composes and fuels them. Behind these productions, in fact, there is a history, and generations of craftsmen who have preserved, renewed, and transferred their knowledge, thereby creating an inimitable collective heritage. This heritage must not be lost. It should be exploited and developed with every available means. It follows that a competitive force of Loro Piana and Gucci lies in the integration of manual skills and technology. A brand that dialogues with sustainability should become an expression of this *integration* fusing in its identity the *values of tradition* with those of *modernity*. All this particularly marks Gucci brand. The sustainable brand, then, is the offspring of a modernity, which does not deny but rather is composed of a past, which strives to become eternally present.

Sustainable luxury includes, finally, a *political* and *institutional dimension* (Fig. 1). Thus, it embeds a propensity towards the *"polis"*, that is towards the acquisition of a space in the governmental decisions of a country. Especially in Loro Piana, Gucci, and Zegna in fact, we are witnessing a search for legitimacy on the part of local institutions. This legitimacy can act a priori or a posteriori. In other words, it can activate a company's sustainable processes or it can formalize their actual existence. The form used consists of agreements signed with local and foreign institutions, such as governments, ministries, and business associations that pursue a socio-environmentally based country ethic. Sustainable luxury has, then, a *diplomatic soul* and even thrives on *formalism* and *acts* that enshrine its existence socially. The recognition of consumers, employees, and suppliers seems not to be enough. Sustainable luxury, in fact, assumes full semblance if integrated in social policies pursuing the collective well-being. Thus, it expands the communities with which it interacts and of which it is the guardian; it becomes visible in a wider, but heterogeneous community, in that of the citizens. Thus, the dialogue between brand values and sustainable luxury may become more symbiotic and interpenetrating to the extent to which it passes through a process of, one could say, *social institutionalization*: the ethical identity of a brand and its sustainability-based actions acquire an intrinsic coherence once they become formally a collective knowledge. This process particularly joins Gucci and Zegna. The social brand value is then reinforced by the political brand value. It is an emerging relation, but potentially also critical. The process of institutionalization, if instrumental to a recognition stabilizing the social brand value or to the activation of cooperative interpersonal exchanges, can create stagnation and disillusionment. If the market, and therefore consumers, became aware of an underlying false truth, they would consider the dialogue between brand values and ethics a mere illusion, and perhaps even a sort of deception. This would contribute to the destruction of the ethical path that a brand has blazed over time. Nonetheless, all the sustainable luxury dimensions investigated, once adequately combined, produce positive effects on brand performances.

5 Conclusions

The case analysis reveals the sources of an interpenetrating relation between luxury brands and sustainability. This relation implies a symbiotic alignment that the brand seeks between its values and the sustainable actions undertaken. The alignment requires a corporate ideology based on quality, excellence, and rigour. Moreover, it passes through a collective sharing within community contexts of the combination of brand values and sustainable projects. The resulting contamination increases mutual respect and collaboration. Then, the materialization of sustainable projects as synthesis of the brand ethics requires craftsmanship and, thus, the flexibility necessary to shape productions generated by progressive technologies. Finally, the dialogue a luxury brand develops with sustainable choices, while it may be legitimated by institutions through formal acts, must remain fluid, be free, and have an independent life; it has not to bureaucratize the underlying businesses processes that are largely inspired by rhythms, not only economic, but also natural. From the analysis of the dimensions of sustainable luxury, it emerges that not only scarcity and durability join luxury and sustainability, but also the distinguishing qualities of rigour, perfection, sharing, flexibility, and independence. Just as nature thrives on a nearly spontaneous rigour, luxury is the result of rules oriented towards the quest for excellence. The natural environment is an open and accessible environment; similarly, luxury is shared by those who want to recognize it and make it part of their existence. The natural world is flexible and manages to reorganize itself even in the wake of catastrophic events; in the same way, luxury comes to life if it is grounded in the manual flexibility of those who are capable of creating it. Finally, the environment is a free collective good; luxury can acquire independence as well, when it is transformed into a social myth, recognized by the broad community. However, despite this commonality, the balance between brand values, ideology, and actions behind a real sustainable luxury is rather difficult to maintain over time. Takeovers, company expansions, and new entrepreneurial generations can denature the pursued sustainable managerial approach. This risk marks all the different cases we have investigated. It follows that we may be suspicious of the genuineness of the sustainable behaviours companies adopt. While these behaviours are surely a priority for many luxury fashion companies, it is possible they hide a zero-sum game with the market. This means that what a company achieves by sustainable actions joined with its perceived brand value may appear much greater than what it has invested in carrying out those actions. Disproportionate gaps, especially in cases of high brand awareness, can reflect a mere economic exploitation of sustainable actions. There emerges, therefore, a twofold imbalance between environment and business: not only economic, but also ethical. Measuring this imbalance is important. This is because a revealed economic imbalance, if it is appropriately considered, could correct the underlying ethical imbalance. The company should reallocate the economic resources it dedicates to sustainable actions in relation to the impact these actions have produced on its brand performance. As a consequence, a declared sustainable responsibility can find its demonstration in an incremental sustainable behaviour based on the achieved eco-performance. Evaluating the existing equality in the relation between luxury brand and the environment can constitute, in our view, a possible path for scientific research within the context of sustainable luxury brands. In the next phases of our research, we intend to follow this path. Fully aware of the limits of an exploratory study concentrated on emblematic cases, and exclusively Italian, we propose to refine our qualitative research to explore even more in detail the facets of the luxury brand dimensions identifying them in all their contents; then, we intend to carry out quantitative research to investigate the impact of each dimension on brand performance. The ultimate objective is to develop specific ratios capable of evaluating the compatibility between the brand performance and the sustainable actions undertaken. We think that if on the one hand, it is difficult to assess thoroughly how sustainable a company is; on the other hand, we can at least estimate to what extent its relationship with the environment is not inspired by temporary, opportunistic, and speculative aims. Understanding the "how" of a luxury company's sustainability is, in our view, ultimately more important than bowing only to its declared and perceived "how much".

References

Achabou MA, Dekhili S (2013) Luxury and sustainable development: is there a match? J Bus Res 66(10):1896–1903. doi:10.1016/jbusres.2013.02.011

Bendell J, Kleanthous A (2007) Deeper luxury: quality and style when the world matters'. www.wwf.org.uk/deeperluxury. Accessed 11 Feb 2015

Berns M, Townend A, Khayat Z, Balagopal B, Reeves M, Hopkins M, Kruschwitz N (2009) The business of sustainability: what it means to managers now. Sloan Manag Rev 51(I):20–26

Closs DJ, Speier C, Meacham N (2011) Sustainability to support end-to-end value chains: the role of supply chain management. J Acad Mark Sci 39(1):101–116. doi:10.1007/s11747-010-0207-4

Colbert BA, Kurucz EC (2007) Three conceptions of triple bottom line business sustainability and the role for HRM. HR, Hum Res Plan 30(1):21–29

Crawford CB, Ranfagni S, Guercini S (2014) Exploring brand associations: an innovative methodological approach. European J Market 48(5/6). doi:http://dx.doi.org/10.1108/EJM-12-2011-0770

Davies IA, Lee Z, Ahonkai I (2012) Do consumers care about ethical luxury? J Bus Ethics 106(1):37–51. doi:10.1007/s10551-011-1071-y

Gardetti MA, Girón ME (2014) Sustainable luxury and social entrepreneurship: stories from the pioneers. Greenleaf Publishing, Sheffield

Guercini S (1996) Considerazioni di metodo sull'impiego della case analysis nel campo degli studi d'impresa. Working paper No. 2, Department of Business Science, University of Florence

Guercini S (2014) New qualitative research methodologies in management. Manag Decis 52(4):662–674. doi:10.1108/MD-11-2013-0592

Guercini S, Ranfagni S (2013) Sustainability and luxury. The Italian case of a supply chain based on native wool. J Corp Citiz 52:76–89. doi:10.9774/GLEAF.4700.2013.de.00008

Hennigs N, Wiedmann KP, Klarmann C, Behrenss S (2013) Sustainability as part of the luxury essence. Delivering value through social and environmental excellence. J Corp Citizensh 52:25–35. doi:10.9774/GLEAF.4700.2013.de.00005

Hult GTM (2011) Market-focused sustainability: market orientation plus! J Acad Mark Sci 39:1–6. doi:10.1007/s11747-010-0223-4

Jones P, Clarke-Hill C, Comfort D, Hillier D (2008) Marketing and sustainability. Market Intell Plann 26(2):123–130. doi:10.1108/02634500810860584

Kapferer JN (1997) Managing luxury brands. J Brand Manag 4:251–260. doi:10.1057/bm.1997.4

Kapferer JM, Michaut-Denizeau A (2015) Are luxury purchasers insensitive to sustainable development? New insights from research. In: Gardetti MA, Torres AL (eds) Sustainable luxury. Greenleaf Publishing Limited, Sheffield

Lubin D, Esty E (2010) The sustainability imperative. Harvard Bus Rev 88(5):42–50

Montiel I (2008) Corporate social responsibility and corporate sustainability: separate pasts, common futures. Organ Environ 21(3):245–269

Pantzalis I (1995) Exclusivity strategies in pricing and brand extension, unpublished doctoral dissertation. University of Arizona, Tucson

Pullman ME, Meloni MJ, Carter CR (2009) Food for thought: social versus environmental sustainability practices and performance outcomes. J Supply Chain Manag 45:38–55. doi:10.1111/j.1745-493X.2009.03175.x

Ranfagni S (2012) Decentramento produttivo e processi di internazionalizzazione. Casi di imprese di abbigliamento. Rubbettino Editore, Cosenza

Ranfagni S, Runfola A (2012) Internazionalizzazione e strategie di marca. Casi d'impresa a confronto nel sistema moda italiano. Finanza, Marketing e Produzione XXX(2):147–176

Rastier F, Cavazza M, Abeillé A (2002) Semantics for descriptions. CSLI, Stanford

Sharma A, Iyer RG, Mehrotra A, Krishnan R (2010) Sustainability and business-to-business marketing: a framework and implication. Ind Mar Manage 39:330–341. doi:10.1016/j.indmarman.2008.11.005

Stewart DW, Kamins MA (1993) Secondary research: information sources and methods. Sage, Newbury Park

Vigneron F, Johnson LW (2004) Measuring perception of brand luxury. J Brand Manag 11:484–506. doi:10.1057/palgrave.bm.2540194(6)

Yin RK (1984) Case study research: design and methods. Sage, Thousand Oaks

Irreplaceable Luxury Garments

Creating Emotional Engagement

Susanne Guldager

Abstract This chapter addresses the value of emotional engagement and the state of irreplaceability of luxury garments, as one way of approaching the topic of sustainable luxury fashion. Engaging with the consumer on an emotional level initiates a strong bond between the garment and wearer, thus increasing the longevity of the garment, which in turn becomes irreplaceable to the user. The fundamental philosophy behind this approach to sustainability is to produce and consume less and better products, and to design products from the idea of doing 'more *in* less'. To accomplish this, time and human presence are the key elements. The value of a luxury garment ought to depend on the following: the idea and the thoughts behind creating the product, the time spent producing the product, focusing on the way it is produced and by whom, and the long-term meaning it will bring to the life of the user. This genesis story has to be tangible and subsequently mediated to the consumer and more importantly experienced and appreciated by him or her as a valuable element of luxury. The concept of luxury fashion needs to be broadened and deepened from something that is widely related to status and image, toward something that concerns us in a deeper (emotional) sense.

Keywords Sustainable luxury processes · Luxury and consumption · Engagement consumer

1 Introduction

The conventional fashion industry is mainly building upon short-term systems of mass-produced fast fashion, in which economy is the key holder to control the market. The decisive competitive parameter is price, and as a result, consumers are highly price focused; thus, they assume that the cheaper a garment is the better.

S. Guldager (✉)
Copenhagen School of Design and Technology, Copenhagen, Denmark
e-mail: sug@kea.dk

© Springer Science+Business Media Singapore 2016
M.A. Gardetti and S.S. Muthu (eds.), *Handbook of Sustainable Luxury Textiles and Fashion*, Environmental Footprints and Eco-design of Products and Processes, DOI 10.1007/978-981-287-742-0_5

Consumption in this way requires a human and moral distinction as to how garments are actually made and the circumstances hereof. On the face of it, this seems as the antithesis of luxury fashion; nevertheless, this industry is neither working for nor against this human and moral distinction. It seems to be a recurring theme in the fashion industry, luxury or not. A bridge is needed to reconnect human agents in an industry characterized by fragmentation, domination, and subdivision of stakeholders concerned with their own interests and agendas.

In extension, Stuart Walker[1] claims that there has been a detachment from our material world, and he addresses:

> A reassessment of physical products is required, together with a creative re-engagement with 'things', if we are to find lasting meaning and value in our material world whilst simultaneously alleviating the damaging consequences of contemporary consumerism[2]

This quote heads against what is advocated for in this chapter, which can be condensed to:

> Re-connecting with the world of luxury fashion by understanding and sustaining the value of time and human presence, and from here aiming for garments to become irreplaceable to the human being by a strong emotional bond.

This will lead to the sustainable goal of creating product-longevity and ultimately to a change or at least an awareness in our behaviorism from consuming 'mass *to* less'. It is important to stress that the present intention is not about assimilating to asceticism and pure anti-consumption. Instead, the intention is to slow down and change focus; to put it philosophically show the need to think about *being* instead of *becoming*—without degrading development processes and the enrichment of transformation. Finally, it is worth noting that this is just one way of addressing the issues of sustainability contextualized in luxury fashion.

1.1 Changing Focus—Doing 'More in Less'

A sustainable industry must be an agent for searching for 'the good,' instead of having to seek 'the new' as a means to an end. This is, however, a huge challenge in a society, dominated by speed and consumption, short-term solutions, and a lack of willingness to engage. As John Thackara[3] addresses, it is a society in which speed and constant acceleration is worshipped uncritically as a generator of investment and innovation.[4]

---

[1]Professor of design for sustainability.
[2]Walker (2006).
[3]Author and founder and director of the event *The Doors of Perception*.
[4]Thackara (2006).

The conventional way of perceiving a company's success is through resource efficiency and to do 'more *with* less'. This also includes reducing the time spent in the development and production process, the amount of human workers to carry out the work, and the volume of raw material used in the production. A reduction in the material resources is, however, a positive way of optimizing production and in correspondence with a discourse of sustainability.

A focus on a success criterion such as doing more *in* less is needed; there is a necessity to reduce the quantities and upgrade the quality in the products being brought to life by increasing the resources concerning time and human presence. In this way ideally, the amount of human workers needed to carry out the production will not be altered, because instead of producing quantity they will spend the same time producing quality. This is essentially consistent with the way many luxury products are already being designed and produced today; yet, there is still place for improvement. To reach the state of irreplaceability also calls for a reassessment and an expansion in our understanding and perception of luxury garments, e.g., by assessing craftsmanship as a luxury. By doing so, the products might, to a greater extent, entail the preconditions for creating an emotional relationship.

1.2 The Emotional Relationship

The emotional relationship to the material world is a source of wonder—why are some products discarded without the slightest consideration, while others are never discarded? Long-lasting emotional attachments to certain garments are formed, but what determines these emotions, how can understanding these emotions help in creating garments from a sustainable point of view? Could designers aim at creating products that activate this 'emotional capital,' and based on this capital could they generate product-longevity? Ideally, this could lead to more people creating a wardrobe, which lasts, because the garments become irreplaceable due to the emotional connection between garment and user. Emotions can impact and effect sustainability, if well managed and utilized in a morally responsible way.

1.3 Methodology

As a foundation, the section '*Luxury Fashion and Sustainability*' will briefly discuss the concept of sustainable luxury fashion, and the contradictions and obstacles as well as the possibilities that lie within the concept. The section that follows, '*The Meaningful Relationship Between Objects and Human Beings,*' elaborates on the mutual interaction that is taking place between human beings and the world of materiality, including design products and moreover luxury garments. How objects in general can be carriers of meaning added by human beings, and mutually how they are creating meaning in lives, affecting behavior and contributing to the process of forming individuals is also explored.

The initial discussion of sustainable luxury fashion and the relationship between objects and human beings constitutes the foundation to deal with the emotional relationship between a user and a luxury garment contextualized in sustainability; therefore, it is important to address these issues before focusing on the *emotional* part of it.

The issue of the section, '*Creating an Emotional Bond,*' covers the importance of emotions in a particular context consisting of the 'emotional capital' of a human being as well as the emotional experience, and an introduction to six different states of emotional engagement. Subsequently, the fifth section presents a framework on the user's relationship to design products as well as the lifetime of such. This framework might be useful as a strategic foundation when designing garments aimed at emancipating emotional and durable relationships.

In the sixth section, '*Reframing Luxury Fashion Through Time and Human Presence,*' the synthesis of the concept of emotional engagement, the philosophy of doing more *in* less, and luxury garments as the objective will be clarified. The point in this section is that the fashion industry as well as the common consumer must deepen and broaden an understanding and perception of luxury garments, e.g., by assessing *originality* and *differences* as a luxury. The basis of influence in attaining this sustainable goal (creating product-longevity and reaching the state of irreplaceability through a strong emotional bond) is by doing more *in* less. Basically, adding time and human presence to the products does this, and four different categories will be introduced and bridged with the aforementioned states of emotional engagement. Furthermore, the significance of telling the story behind such garments is addressed, and also a suggestion on how to articulate the consumer of these (new) luxury garments. Finally, the chapter presents a brief summary of the most important points introduced.

2 Luxury Fashion and Sustainability

Luxury fashion is often associated with embellishment, beautification, and expensive habits, and in many contexts, it is a way of displaying beauty, youth, power, and wealth, which leads to recognition from other people. An interesting aspect is the major impact that a brand name has in these social manifestations. In many, maybe even the majority of luxury garments, the brand value is of such a high significance that it per se constitutes the main purpose of possessing the items. Furthermore, much luxury fashion is subject to dynamics based on continual changes and the elusiveness of fashion, which are dynamics supported by replacement logics. In spite of this, there are still many luxury garments that are relatively enduring over time, because of the craft, the exclusive fabrics being used in the making and the cultural heritage of a luxury company.

When isolating fashion as a concept, it is often understood as something that relates to clothing, i.e., as a set of 'rules' for how one should dress. However, fashion can have many facets and interface within numerous disciplines. Fashion can be defined as seeking the new and by the rapidly changing guidelines for how to think and act in a particular cultural community at a given time,[5] and in some instances, it even influences body language, behavior in social situations, and how one linguistically expresses oneself. According to Lars Frederik Händler Svendsen,[6] fashion is neither a universal phenomenon that exists everywhere and at all times, nor something that is a part of human nature or group mechanisms in general. Fashion is a social-constructed phenomenon that has occurred in a society, and with time spreads to other societies[7], and that is a dynamic, which has great impact on the individual as well as on a collective and societal level, reflected in the span between how one should look, to how one organizes one's life. According to Svendsen the essence of fashion is to be fleeting.[8]

By cultivating the new, fashion garments must be produced in a constant flow, and this is basically contradictory to a modern way of sustainable thinking. Fashion critic Vanessa Friedman highlighted the dilemma of sustainable fashion during Copenhagen Fashion Summit in 2014. Friedman claimed that:

(…) fashion is onetime-oriented.[9]

She refined the problem of sustainable fashion by stating that on the one hand, there is a pressure to constantly follow new trends, and on the other hand, there is an indisputable need for retention.

Nevertheless, it is worth noting that there are numerous approaches in dealing with sustainability in the context of fashion, including luxury fashion, and that is an important point when investigating and articulating the concept of sustainable fashion. Sustainability is a way of thinking, which represents an enormous potential that must not be limited by one particular belief such as: sustainable fashion is about using, ecological cotton or recycled polyester e.g.—this could restrict creative thinking, which is invaluable when it comes to integrating sustainability as a natural part of our thinking and behavior.

As a sender of a sustainable luxury garment, a company can have many good intentions regarding the production, in relation to environmental considerations, social circumstances, by using recyclable or disposable materials. This is by no means a guarantee that the garment will continue on a sustainable course, when it begins its 'life' in the hands of a particular consumer. To justify a luxury garment as morally and ethically sustainable due to its production methods and the circumstances hereof, it ought to be an indisputable and simultaneous endeavor to extend

---

[5]Den danske ordbog—moderne dansk sprog (2015).
[6]Norwegian philosopher and author.
[7]Svendsen (2005).
[8]Svendsen (2005).
[9]Holbech (2014).

and prolong the lifetime of such a garment once it is in the hands of the end user. This is one way of dealing with the complexity of sustainability, as well as a holistic approach.

As previously mentioned, luxury fashion already fulfills this requirement to a great extent, for instance, due to the widespread durability of such products, yet this does not mean that they are irreplaceable to the user. Of course, such products often have a monetary value, but still he or she would not hesitate to replace the product if necessary or even just possible. As a response and counterpoint to this replacement logic, the emotional state of irreplaceability becomes interesting. Irreplaceability arises the moment there is a comprehensive engagement and commitment based on emotions. Thus, the way things are made and not least how one emotional relates to them in the use phase will be the main focus points for the subsequent discussion. As a foundation for this, the human relationship to the world of materiality and the power of design will be elaborated upon in the following section.

3 The Meaningful Relationship Between Objects and Human Beings

Objects are a part of forming one as a human being. Through temporal and spatial passages, objects are anchored in the individual and the world. Also, the relationships to objects are interconnected with social processes and influences of how human beings act and behave. Basically, one's understanding of oneself and the surrounding world is molded through dealing and interacting with everyday objects.[10]

The human relationship to objects can be contemplated from two perspectives:

> 1) The human being confers value and meaning to the object whereby this becomes a carrier of these properties.2) The object influences the human being by creating meaning in his or her life, often in terms of identity formation.[11]

Thus, there is a kind of mutual and reciprocal interaction between the material world of objects and the sentient, perceptive human being—this often perceived as something that takes place on an unconscious and non-articulated level.

3.1 Intentionality Regarding Material Possessions

Consumers have an intentionality concerning materiality and consumption because through material belongings, subjectivity and individuality arises. The garments

---

[10]Kragelund and Otto (2005).
[11]Kragelund and Otto (2005).

worn by an individual serve as a medium for communicating with the people around him or her and nonverbally tell something about his or hers individual character and peculiarities. This intangible and unspoken communication is based on culturally constituted sign systems and is understood in the context of other people, objects and situations. This sort of communication is an enriching process in defining 'you' to the surroundings.

Additionally, each human being has an individual 'filter' through which products are perceived, assessed, and experienced. Therefore, as a sender of a product, it is crucial to have a thorough knowledge and understanding of the consumer regarding his or her needs, positions, preferences, ideologies, beliefs, social, and cultural circumstances. This is important when aiming to make people decode a product a specific way, attach certain values to it, and awaken particular feelings in the consumer experience and interaction with the product. Inherently, as a fashion designer, this is of great importance in the attempt at creating garments that touch the consumer on an emotional level and furthermore in the process of generating an emotional relationship between the human being and the product.

3.2 Design for Change

Regarding the reciprocal relationship between objects and human beings, physical products have the ability to affect behavior, and interfere with the way an individual conducts his or hers life—just think of the smartphone. As Thackara writes in his book: *In the Bubble—Designing in a Complex World:*

If we can design our way into difficulty, we can design our way out.[12]

From this perspective, design products and also design thinking in general can be key components in mediating and evolving positive change regarding one's way of thinking about and being aware of consumption patterns. It might even be a central part of creating the foundation for a paradigm shift concerning sustainability.

As a concept, *design* is well known and used broadly in many situations and by many people. Also, it is a word that is used both as a verb that refers to the act of designing and as a noun that refers to a product as well as a process.[13] In the quote below, John Heskett[14] addresses the complexity of the word:

Design is to design a design to produce a design.[15]

---

[12]Thackara (2006).
[13]Lawson (2005).
[14]Author and professor in design.
[15]Heskett (2001).

However, design is often described as a functional object that through its appearance (color, composition, material, and structure) constitutes an aesthetic experience—an experience, which is defined and nuanced depending on the subject experiencing. Nevertheless, like sustainability, design is a word full of inconsistencies and embraces innumerable manifestations within today's expanded concept of design, which among others emphasizes design products as physical, material objects as well as immaterial products.

A common denominator is that design products (material or immaterial) reflect and influence tendencies in society and thereby individual lives. Such societal dynamics will always have an impact on the development of products, and this based on the compelling necessity to evolve over time. Conversely, inventive and novel products can generate new emerging tendencies that affect social as well as cultural conditions.

One of many who have defined design is Herbert Alexander Simon.[16] His definition is as follows:

> Everyone designs who devises courses of action aimed at changing existing situations into preferred ones.[17]

This definition emphasizes the power of design, and not least the responsibility that immanently lies within the concept. Consistent with this definition of the concept of design, the preferred situation, and what is advocated for in the current chapter, is designing durable products created with human awareness and presence; quality products, which the end user will engage with on an emotional level, and hence this causes a sustainable approach to consumption in terms of consuming mass to less.

To do this, designers must create products that 'speak' to the emotional capital of human beings, which is a term that will be elaborated on later in the following section.

4 Creating an Emotional Bond

> Design, stripped to its essence, can be defined as the human capacity to shape and make our environment in ways without precedent in nature, to serve our needs and give meaning to our lives.[18]

Value in relation to design is a word with many facets: functionality, symbolism, ideology, emotions, economics, history, sociality, and culture. But what Heskett's

---

[16]Political scientist, economist, sociologist, and psychologist.
[17]Simon (1996).
[18]Heskett (2002).

quote emphasizes is the fact, that design regards human *needs* and brings *meaning* to one's life.

The modernists' way of assessing 'good design' was defined by simplicity, rationalism, logic, and above all practicality. They, indeed, were trying to educate the consumer into being 'good' citizens. Today, this product ideology remains highly visible in a broad spectrum of design products, especially in the Scandinavian countries. What has changed since then is the comprehensive focus on the experience emanating from the interaction between the consumer and the product. The products must be imbued with meaning. They must be aligned with the consumer's needs, based on personal values, to bring some kind of meaning to his or hers everyday life. Thus, a value base of a product controls the consumer's (subjective) meaningful relationship to that specific product.

This concept is of great importance to fashion designers attempting to make garments which endure and coevolve in a long-term time perspective for the user. Also, when a designer manages to establish sensibility and presence in a product, meaning is often emancipated—a kind of meaning, which might touch the consumer on an emotional level and hereby activate his or her emotional capital.

4.1 Involving the 'Emotional Capital'

Emotional capital is typically perceived as a business term used to define the emotional capital in a company. This terminology is based on the employees' psychological assets and resources, such as how they feel about working in that company and their relationship to their colleagues. Emotional capital is beneficial to a company in generating success.

This is, however, not the way the concept will be applied in the present context. Rather the emotional capital of a human being is to be seen more broadly as a human resource that consists of emotions, beliefs, and imagination. It is a resource that many design products capitalize on due to economic benefits—in a worthy as well as a in a less worthy manner.

In paying attention to the emotional part of a design product, it is important to highlight that this is often strategically utilized solely to increase the sales of a product and in some cases even on fake premises. It may concern products that manipulate consumers on an emotional level by playing on fear, a fear, for example, that ultimately leads to the irrational decision that an individual cannot live without this specific product, even though this may not be the actual case.

It might be said that striving to achieve an emotional reaction from the consumer through a product's design is concerned strategic thinking. However, there is a big difference in how this is approached. By exploiting the emotional side of a human being with the main purpose of making him or her buy more, buy irrationally and worst of all, buy on fake conditions, is doing so in an unworthy manner. Instead, to connect with the consumer's emotional resources in order to influence one's purchase of a product that intentionally would emerge as an emotional engagement is another story. One does need to think strategically if one wants to make a change to a more sustainable way of thinking and assessing products. This can, however, be done in an authentic and transparent manner, without further motives other than educating the consumer to choose well, based on a deeply, enduring desire.

4.2 The Value of an Emotional Experience

> All of the experiences you've acquired in your life and work are not sterile facts, but emotionally laden memories stored in the brain. Your life wisdom presents itself as instantaneous hunches and gut feelings… and can dramatically increase accuracy and efficiency of the decision process.[19]

Emotions influence how one feels, experiences, and thinks, but also how one behaves, acts, and makes decisions. Emotional experiences are something that inevitably arise in human beings, and to interact with products that emancipate positive and pleasure-based emotional reactions are something human beings automatically seek.

Thus, the immanent emotional meaning and value in a product not only has to be understood by the consumer, but also more vitally experienced as truly, authentic value-based characteristics. In this perspective, the function of a product is explored as also being a bearer of a story and conveying a deeper emotional meaning to the consumer. This makes value and meaning gatekeepers in activating the emotional capital and the means necessary to liberate an emotional experience.

In a design perspective, emotional experiences are of great value, and if the 'language' of a design product 'speak' to the consumer on an emotional level, it might lead to a purchase, and this may also affect the longevity of such a product. Donald Norman[20] supports this perspective by stating:

> I argue that the emotional side of design may be more critical to a product's success that its practical elements.[21]

Thus, emotions have to be taken into account when dealing with design products, including luxury garments, as they have major impact on our human behavior, and in the actual discussion more vitally when we relate to them in the actual use. Donald Norman continues:

> In creating a product, a designer has many factors to consider: the choice of material, the manufacturing method, the way the product is marketed, cost and practicality, and how easy the product is to use, to understand. But what many people don't realize is that there is also a strong emotional component to how products are designed and put to use.[22]

---

[19]Cooper and Sawaf (1997).
[20]Engineer and mathematical psychologist.
[21]Norman (2005).
[22]Norman (2005).

The emotional experience is the key component when discussing durability in terms of commitment. It can be activated in the moment of the actual purchase situation, because of the ability of a product to appeal to and awaken immediate feelings such as empathy, nostalgia, and happiness. In this preliminary meeting, the real commitment is still not definite and existing, but the root hereof is brought to life. The genuine commitment in most cases emerges over time and by sustained use.

4.3 The Nuances of Emotional Engagement

Emotional engagement and commitment are based on a product's ability to cultivate long-lasting feelings in a human being and by meeting the human needs and desires during time. There are several ways of applying emotional value and awakening an emotional engagement. Below are six suggestions:

Empathy and intimacy
Nostalgia
Self-trust and self-confidence
Morality
Sympathy and love
Personalization and rediscovering

Highlighting the craftsmanship or the maker of a garment might lead to a kind of *empathy* and *intimacy*, from the consumer. Evoking these feelings could be an essential part of triggering an emotional engagement to a garment. If a designer manages to activate the consumer's emotional capital and foster some kind of empathic or intimate feeling within him or her, the designer already covers a great distance toward the objective of creating an emotional engagement.

Equally, the value of *nostalgia* is able to stimulate the emotional capital of a human being. Everyone knows the feeling of being reminded of his or her childhood, perhaps through memories of a grandfather's home-knitted sweater or of a moms' seventies leather bag that had that specific smell, memories which arouse pleasurable feelings.

Moreover, the emotional engagement to a garment could be supported in desirable feelings such as *self-trust* and *self-confidence:* wearing a coat with a fit that emphasizes the body shape in a positive manner, or a pair of trousers that liberates the feeling of pure comfort, or to put on a shirt with a remarkable expression that makes one stand out of the crowd.

The conditions of emotional engagement can furthermore be of a *moral* character where the user feels a responsibility, obligation, and commitment to take care of the garment through the self-felt obligation of repairing and gently maintaining it, which can be labeled an 'emotional investment'. If a state is reached, where the user wants to repair such a garment instead of throwing it away, some kind of emotional connectedness has arisen between the user and the garment.

The emotional state could also be characterized as *sympathetic* and *loving* equivalent to the relationship to an animal or between human beings. In some cases, the relationship can even be perceived as a symbiosis between the human being and the garment—a mutual process where the user transmits emotions to the garment and the garment somehow affects the user on a deeply emotional level. This emotional state, though, is rarely initiated in the very beginning of a relationship, but is usually developed over time.

1. Anecdote—My beloved boots

> I recognize the feeling of emotional engagement with a pair of my own brown – and very beloved – leather boots. I clearly recall the purchase-moment four years ago. I memorize thinking that these boots were quality boots regarding the material, the fit and the color. From this quality-perspective I argued then, that the price was worth paying due to the longevity of the product.
>
> As predicted this actually was a product of great longevity, and during time a feeling of emotional connectedness has arisen, and I cannot image ever to get rid of them. After years of intensive use – in spite of the quality – one of the zipper-heads has fallen of. In the beginning it could be replaced with a little metal-thing from my key-holder. Unfortunately, the zipper is too broken for this method now, but I can still zip and unzip the boot by using my key, and that does not affect how often I bother to wear them.
>
> The only time I regretted wearing them was in the airport of Colombo in Sri Lanka, when the security guard asked me to take of my boots, and I had already left my bag, including my key, on the security band. The woman in front of me started laughing, when I tried to explain the situation, and the security guard let me through – with my boot still on.
>
> An additional aspect of this anecdote is, that at the time, when I had that little metal-thing attached to make it out for the zipper-head, my love and engagement to the boots simply increased. Suddenly I had added my own personal character to the boot, and by doing so I made them unique.[23]

Personalization of belongings is an interesting part of emotional engagement. By doing so, there is no existing copy, which is completely similar. This uniqueness is in some instances the actual reason for not wanting to discard such a product, and exactly this feeling of irreplaceability is the goal in the actual discussion about emotional commitment and engagement.

A designer might chose to let a space remain open for the user to *personalize* or being able to *rediscover* the product, and to add his or her individual character to the overall expression of the product. Such an action makes the user a cocreator of the product and must not be underestimated in the establishment of an emotional connection. This might also create a substitute for the desire to buy something new, cf. that fashion is often associated with searching the new.

The six above-mentioned aspects are just some types of emotional engagement, but what they all have in common is their ability to initiate a strong bond between the user and a specific garment, thus bringing a long-lasting meaning to that person's life. This is definitely a noteworthy way of meeting the need of promoting and

---

[23]A personal anecdote from the author's life.

actively dealing with sustainability, an awareness also addressed in following quote by Walker:

A sustainable solution can be understood as one that possesses enduring value in terms of its meanings and characteristics.[24]

An important issue concerning emotional value and meaning, which constitute the foundation of emotional engagement, is that these values and meanings are non-static attributes when experienced in relation to a product. Value and meaning is determined by time, space, and situation and must be considered as dynamic entities. One's relationship to products is submitted to constant change. Some relationships become more valuable and precious over time due to an emotional attachment based on subjective memories related to situations and people from that specific person's life.

Obviously, emotional engagement is based on highly subjective matters, which makes it hard to objectify and strategically to target as a designer of a product. Nonetheless, this engagement can be supported in many ways, for instance, by communicating the story of the product life, by encouraging the user to respect and take care of the product, and to set the stage for customizing and personalization. It is also worth noting that sustained use is a vital part of creating the basis for an emergent dynamic relationship that might move toward an emotional engagement. Thus, as a sender of a product, it is crucial to produce garments, which can withstand use—more than twice. This leads to the next section, which introduces a framework to be used to articulate two different systems of a product's life cycle, based on the human relationship to it.

5 A Framework for Relationships to Garments

The framework (illustrated in Fig. 1) visualizes two different cycles of a product's life. Each of these distinct cycles represents, respectively, a short-term and a long-term system based on the relationship to and the lifetime of a product. To clarify one system from the other, they will be simplified and mainly articulated as each other's opposites. Although the framework will be verbalized in the context of (luxury) garments, it can also be applied to other product categories within the field of design.

5.1 The Short-Term System

The lower circle of the framework suggests a system consisting of short-term solutions when addressing consumption issues. It classifies products that most

---

[24]Walker (2006).

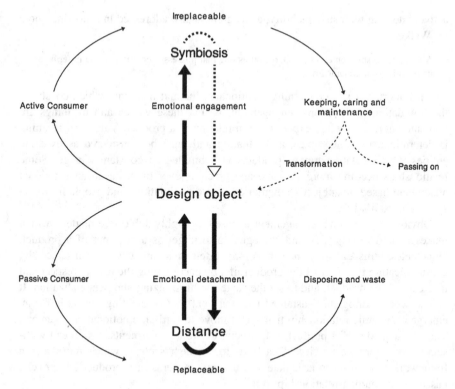

Fig. 1 Framework for relationships to design objects and the life cycle of these

likely will have a short lifetime or at least a short-term value and meaning for the user. A system, nevertheless, that encompasses most of today's products.

Object character	The focus is purely on the product's material properties being used as a utilitarian tool. These kinds of objects can be seen as 'bodily tools.' The use consists of physical elements such as efficiency and convenience, e.g., to cover a body or to keep it warm, but it also consists of the use of social, cultural, and societal dynamics. This can be found through the dissemination of signs and symbols, which support the manifestation of the user's identity to the surrounding social world. The attention is directed toward how these garments can help a user in a pragmatic perspective, consciously as well as unconsciously.
User character	The user could be characterized as passive, because (roughly speaking) he or she feels separated and detached from the products, which are to be perceived as purely objective articles.

| Relationship character | In relationships between garments and the user, there is no profound emotional connection, nor does it add a deeper meaning to his or her life. Thus, many such relationships could be classified as distanced, in different degrees. |
| Lifetime character | Because of the detachment and noninvolvement of the user, there is a greater tendency toward such products being easy to replace and can be disposed of without further consideration or emotional response. |

Garments like these do not touch the user on an emotional level, and a consequence of this is that he or she does not want to make any effort into repairing them —in some cases, the user might not even consider this as a possibility, and in other cases, he or she simply does not want to pay the costs of a repair, maybe because it is often cheaper to buy a new one.

5.2 The Long-Term System

The upper circle of the framework visualizes a long-term system in which the focus is on how the user relates emotionally to and engages with products, rather than what the product can impart to a user in a pragmatic short-term sense.

Object character	Garments in this system also respond to a practical need, but they are primarily to be portrayed as 'mental tools' that somehow represent a deeper meaning in the lives of the users.
User character	The user is to be seen as active and engaged; he or she experiences an enhanced feeling of emotional enrichment.
Relationship character	The relationship emphasizes a deeper meaning, which is established by an emotional involvement. An intimacy and connectedness is anchoring the bond between the user and the product, and this groundwork becomes a fertile soil for a long-term emotional engagement. The object and the user mutually constitute each other, and in some cases, the relationship would even be characterized as symbiotic.
Lifetime character	There is a willingness to keep, care for and maintain such a garment, and because of the emotional bond, the product obtains a character of irreplaceability. In case the user gets tired of the garment, he or she might transform it into a renewed product, otherwise get rid of it by passing it on to a beloved or by selling it. Either way it will keep circulating and thereby stay part of a sustainable (long-term) system.

Emotional engagement is a venue that embraces both the consumer as well as the product, and the focus is increasingly on what occurs in the space between these two parts. This emotional atmospheric space is vital to support the shift from consuming mass to less. For this to be realized, the designer and the different stakeholders need to set the stage for adding more *in* less, and for the user to coevolve and coexist with the product during time.

5.3 Dynamic Movements

> Time involves a kind of movement or activity. It does not stand still. It waits for no man. Sometimes it even flies. Poets liken time to a river, bringing fresh events and sweeping away old ones. Time is always passing. (...) And things get older: even if they don't wear out or lose their hair or change in any other way, their chronological age is always increasing. These changes are universal and inescapable: no event could ever fail to be first future, then present, then past, and no persisting thing can avoid growing older. We call this process time's passage.[25]

Durability has already been mentioned several times and is inevitable when discussing emotional engagement. These two concepts, durability and emotional engagement, are inseparable and act simultaneously when one reviews the issues of the current debate. If a user is emotionally engaged to a garment due to different circumstances, his or her willingness to keep that garment increases, and on this basis, durability arises. In the attempt of establishing an emotional connection, the fact of possessing a garment for a long time is of abundant value; the emotional bond between a user and a product becomes more solid in the passage of time, due to the accumulation of 'narrative' layers. This makes it a coexisting and reciprocal interaction where emotional conditions and durability are interconnected.

Due to the fact that emotional engagement is extremely elusive and subjective, it is difficult to predict what triggers it, and as mentioned previously, relationships to products are to be seen as dynamic, constantly shifting and transformative; garments express temporal and spatial changes and transitions from one state to another and become metaphorical evidence of time and memory. It is likely that the character of a relationship to such garment mutates, so to speak, during time.

Hence, occasionally a garment, which basically has been purchased as a 'bodily tool,' and is hereby allegedly part of the short-term system, transfers to the long-term system during time. The former distance in the relationship condenses, and a higher level of proximity emerges. Ezio Manzini[26] addresses this developing process in a poetic way:

---

[25]Olsen (2009).
[26]Industrial ecologist.

To appreciate quality, I have to take time. With a glass of wine I have to smell it, look at it, I have to take my time to drink this wine. Even beyond that, to be able to understand that is a good glass of wine, I had to do something before – to learn, to spend time in study.[27]

In contrast there is a greater inertia in the diverse direction; it appears rather unlikely that a user becomes emotional detached from a garment that he or she has already experienced a strong emotional bond and connection with, and yet it is not an impossible occurrence.

5.4 Luxury Fashion Contextualized in the Framework

Luxury garments do not represent the majority of articles in the short-term system. If a user has bought a luxury garment, it supposedly has a high monetary value, which would probably awaken some kind of intrinsic resistance against throwing such garments out. For this reason and because of the high quality they possess, most of such items fall into the long-term system. However, this does not implicitly make the product irreplaceable, which in a sustainable manner is the main goal of emotional engagement. If a luxury garment for instance breaks, and you are not emotionally connected to that specific garment, you would not feel any (emotional) resistance in replacing it.

As previously mentioned, it is difficult to predict whether a user can become emotionally connected and engaged to a garment or not. In the section '4.3. The Nuances of Emotional Engagement,' six different proposals of emotional engagement were introduced. These proposals referred to the emotions: empathy and intimacy, nostalgia, self-trust and self-confidence, morality, sympathy and love, as well as personalization and rediscovering. Yet, to integrate some of these emotions into the preliminary work in the making of a luxury garment is not impossible.

Initially, the concept of luxury needs to be deepened and broadened. In the next section, luxury fashion will be reframed as a concept that is referring to more than brand value and monetary value. Time and human presence play key roles in this reframing. The reframing will be classified into four categories, and the afore-mentioned six emotional states will continuously be interwoven as support elements for the emotional connection between the user and the product.

6 Reframing Luxury Fashion Through Time and Human Presence

Stuart Walker claims that there has been a detachment from our material world, due to a shift in:

---

[27]Thackara (2006).

1) The way things are produced; from being created by our own hands to being composed of different components from various countries.2) The quality and quantities of these products; from less products that were built to last to mass products as disposable objects.3) The origin of the production; from local or regional to global. The result of this has led to a lack of understanding and a devaluing of material culture.[28]

Inspired by these three points, the concept of luxury fashion will be reframed in perception and definition. Quality is a vital element in this process and can be expressed in many ways. The topics listed below are concrete suggestions as to how luxury can be reassessed, in terms of following the ideology of doing more *in* less and adding emotional value through time and human presence:

Originality as luxury
Differences as luxury
Sensing as luxury
Slowness as luxury

6.1 Originality as Luxury

Spending time in the initial design process of a garment ought to be appreciated as a quality in itself—obviously if well managed. If diversity is to be kept within the world of fashion, there must be an appreciation of the idea and the concept behind a garment. Elements that have carefully been thought through and developed on an original basis are vital to this process—as opposed to something being copied by others from the catwalk, mass produced and distributed in stores a month later. The latter is a reality that permeates the fashion industry today and that simply increases the substantial value of 'the original'.

The general perception of 'original' is that it has not derived from something else either directly copied or adapted from something already existing, thus conflicting with the way many garments are actually being made, including luxury garments. An ongoing discussion in the fashion industry is whether there is any originality left at all and whether garments nowadays are not merely direct adaptions of contemporary trends or repetitions of fashions from former decades. These being defined as adaptions and repetitions without a genuine autonomous interpretation. To a great extent it is, but it is too radical to write off the entire fashion industry as entirely unoriginal. In this context, it is also important to point out that originality is far more prevalent in luxury fashion than in the conventional fast-fashion industry.

Issues such as adaptable technologies, intelligent materials, 3D printing, or even clothing grown from living organisms embrace originality because of the newness in their methods and technologies. Originality, though, can also be expressed on a far smaller scale based on transforming and putting together familiar elements in a new way and to use the already existing elements in an unexpected manner.

---

[28]Walker (2006).

Generating originality in fashion does not necessarily have to be more complex than spending time examining new combinations and expressions within materials, colors, cuts, silhouettes, and details.

The mindset of original thinking is based upon the ability and courage to break from habitual ideas and assumptions and includes curiosity, wonder, openness, autonomy, and not least an essential interest in creating change. Originality is closely linked to creative thinking. This concept has been the subject of Csikszentmihalyi's (2013)[29] research. He divides creative thinking into two categories: convergent and divergent thinking. Convergent thinking is defined as quickly committing to certain ideas, whose processes are based on a linear development, in which there is an absence of 'the unexpected.' In contrast, divergent thinking is conditional on processes, in which ideas arise diversely and with room for edginess—this way of thinking is the essence of originality. Such thinking takes time and a change in habits, which basically is a scarcity in the fashion industry, and in general is perceived as a luxury by today's standards of living.

Besides cultivating diversity, variety, and multiplicity within the fashion industry, creative processes may also impart a metaphysical connection to the creator; the designer behind a specific garment can impart a sense of humanness in his or her actual design. This aspect needs to be emphasized in an industry dominated by a pronounced distance between the humans involved: the creator, the producer, and the end user; moreover, in a society where human presence, likewise time, is becoming a value in itself, this metaphysical human contact might foster an intrinsic emotional value represented by feelings. These feelings could include empathy and intimacy and could become an integral part in the making of the garment, as is visible in that the garment becomes a medium to awaken the user on an emotional level. Potentially, this process also has the ability to touch the user in a moral sense. When a psychological human distance is reduced and a human presence is activated, a deeper respect for the garment and a willingness to take better care of it are discovered.

6.2 Differences as Luxury

In line with reframing luxury as appreciating originality, there are differences, which can be categorized within luxury. Kate Fletcher[30] and Lynda Grose[31] state that there is a:

> lack of choice and variety of garments on the high street as low-cost 'big-box' retailers create a dynamic that prioritizes cheapness, mass availability and volume purchasing above all.[32]

---

[29]Professor of psychology and management and inventor of the *Flow Theory*.
[30]Researcher, author, consultant and design activist.
[31]Associate professor in fashion design.
[32]Fletcher and Grose (2012).

Thus, garments produced as one-offs and in small quantities ought to be appreciated for their unique character. Product uniqueness is also achieved through craftsmanship, but it can moreover be established if the designer in the design phase of a garment creates opportunities for the user to personalize and rediscover the garment over time during the post-purchase phase.

The small differences and dissimilarities resulting from garments that are personalized or handmade are little treasures in an ocean of homogenized products. Garments that are fully handcrafted or have a craft integrated into their design might indeed be defined as a 'cultural luxury.'

Sensing the maker in a crafted garment is essential when discussing the term of luxury. The paramount industrialization that has characterized the textile industry in several decades plays into this discussion.

> We believed that the assembly line and standardization would make the world a better place, yet along with efficiency came a dehumanization of work.[33]

In distinction to such efficient production, craftsmanship such as hand looming, bobbin lace making, leatherwork, crocheting, and knitting is based on human presence, defined as a skilled-based production method, which requires both time and human resources. According to this definition, craftsmanship derogates from a dominant contemporary trend and manifests itself as the antithesis of a time where everything is meant to go faster and faster, where volatility and boundless replacement constitutes the dominant parameters, that craftsmanship are naturally distinguished from, for a 'healthy' growing society.

The emotional value concerned with uniqueness is united again with empathy and respect as a result of the increased time consumption, and close connection to the actual maker. What supports these emotions is the telling of the story, of the actual making, the process, as well as highlighting by whom it was made. In doing so, the sender of the product shares knowledge and communicates with the consumer and thus educating him or her in a better understanding, appreciation, and evaluation of the resources spent in the making of such a garment.

Additionally, uniqueness stimulates emotions such as self-trust and self-confidence, as the uniqueness of the garment transmits itself to the wearer, as he or she in turn begins to feel special. This is, however, often happening within a secure and comfortable setting. And finally, the emotion of nostalgia is connected to this category through a historical and cultural factor inherent in craftsmanship.

Besides the essential and rare beauty in crafted garments and the emotional value associated with them, the cultural process of maintaining the great heritage based on traditions which have been passed on for generations is of great value. The different ancient techniques and knowledge within these techniques and how they have been demonstrated and transferred to new generations should also be considered. This precious transference is invaluable in the process of obtaining these human-based 'cultural fortunes' that can be defined as articles of luxury.

---

[33]Thackara (2006).

6.3   Sensing as Luxury

The aesthetic sensuous experience of a garment is also worth noting in the discussion of luxury, and in particular fulfilling basic physical needs such as comfort and fit. The pleasurable tactile experience one gets by touching a material or how a garment makes one feel on a psychological level is also noteworthy.

Material quality can be categorized in two ways: 'comfort/remarkable' and 'durable/changeable'. The first component is defined as the comfort and pleasantness experienced by touching and sensing a garment due to its materials. The opposite position within this component is that the material might be less convenient to wear as it is regarded as 'remarkable,' as the quality and the appreciation tied to it emulates a particular expression; however, it is important to mention that the two components are not mutually exclusive and can easily be present simultaneously.

In some cases, these concerns can also be applied to the second category: durable/changeable, even though these two components are slightly more divergent. A material can be of such high quality that it virtually does not change by use; the surface, the stretchiness, and the color fastness remain the same during time. Conversely, a material can also achieve a worn and faded expression through use. This is in general perceived as a lack of quality, but in reality it is not necessarily always the case. The changes that a material undergoes over time can actually be greatly appreciated. For instance, materials such as leather, denim, and cotton often become through an intensified softness more comfortable over time, which increases the emotional pleasure of wearing such garments.

Worshipping the aging process can also relate to an emerging patina that may increase the beauty and the joy of a garment. Also, the symbolic value of such (authentic) altering is not insignificant in the context of the counter reaction to today's throwaway culture. In alignment with this discourse, there seems to be a deeper respect for and appreciation of keeping your clothes until they fall apart.

2.   Anecdote—The man and his boots

> Recently I saw a man wearing a pair of worn-out leather boots, and the front of the right boot was even covered with duct tape. I wanted to ask him why, but hesitated and he was gone. Now I am left guessing as to why. The most rational explanation would be, that he could not afford a new pair of boots. Nevertheless, I got the feeling, that this man really loved his boots, and appreciated the worn expression that they had gained, and simply perceived it as a value of inordinate substance.[34]

Just as the choice of material has a great impact on the quality and the feeling of luxury, so does the comfort of a fit; e.g., how does a dress feel to wear, when interacting with the body, and does a coat follow or counteract the movement of the body? If a garment is to be referred to as an article of sustainable luxury, no matter how bizarre and exaggerated the shape is, this author would argue that it is

---

[34]A personal anecdote from the author's life.

necessary that a garment feels comfortable to wear due to the fit. If that is not the case, maybe such a garment should be defined as an art object, which must to be placed into a different category other than a sustainable luxury garment. The reason being that when a garment is uncomfortable to wear due to the fit, it will most likely end up hanging unused in the wardrobe like a 'virgin garment.' This is not consistent with the intention of this chapter, in which the actual use is an essential part of the concept of emotional engagement.

Cutting, detailing, and finishing also play a big part in assessing garments as luxury. The feeling of wearing a complete and carefully worked-through garment often depends on a cut, a detail and the finish: a beautifully shaped collar, a delicate detail at the pocket, an elegant inner seam covered with a binding, and a lining that harmonize or purposely disharmonize the outer fabric. Such subtleties have great impact on the feeling and perception of quality, and additionally, they are all time-consuming components in the production, which underpin them as elements of luxury.

Ideally, all of the aforementioned aspects ought to be consistently encountered and provide a holistic sensuous experience when perceiving and sensing a luxury garment. Such sensuous experiences, together with a feeling that the garment has been made for and by a human being, may also impart an understanding of the human input as well as the time resources spent in making a thoroughly prepared garment. This is just as plausible in awakening emotions related to empathy and moral, as it is the case with *originality* and *differences*.

Additionally, feelings such as self-trust and self-confidence are connected to the sensuous experience. When wearing a thoroughly prepared garment made in a well-chosen material and with a perfect fit, most likely a feeling of pure pleasure and confidence is initiated. Potentially, during time and intensive use, this might also lead to the emotional state of 'love.'

6.4 Slowness as Luxury

As well as designing people back into the picture, we need to design ourselves more time to paint it.[35]

Time is a recurring theme when luxury is examined in a broadened way. Thackara also emphasizes the fact that lack of time affects the process of observing the consequences of design-related initiatives, as well as our ability to reflect on how the bigger picture is changing.[36]

Doing things quickly implies that we can do more things.[37]

---

[35]Thackara (2006).

[36]Thackara (2006).

[37]Fletcher and Grose (2012).

This basically conflicts with the ideology of doing more *in* less, and it is exactly what the 'slow movement' is opposing. As mentioned previously, products reflect as well as influence the zeitgeist, and these are two processes that mutually affect one another. The slow movement, an existing tendency in some Western cultures, is paradoxically an inescapable fact of life in many less wealthy societies. Nevertheless, it is a tendency that exists parallel with the actual discussion about reframing luxury fashion as multifaceted quality garments, which in various ways encourage longevity based on the manifestation of time and human presence. Also, it reflects a knowledge and understanding of how products are made and why these ought to be appreciated as durable quality products.

When spending more time developing and manufacturing a garment, the quality automatically increases. As Manzini puts it:

Slowness is fundamental to quality.[38]

Slowness as luxury somehow embraces all the different ways of reassessing luxury as mentioned above; from understanding, time spent in the initial creative design process involves thinking and reflecting as a luxury, to perceiving more concrete and practical processes such as tailoring and craftsmanship and small productions as luxury. These processes in their own way are to be stated as luxury of slowness, and something that is worth paying for.

6.5   Telling the Story

There is a slightly different way of articulating luxury other than how the term typically is verbalized. Luxury does not necessarily have to be defined through monetary value, at least that *should not* be what defines and determines it as luxury. However, many of the processes mentioned above require resources mainly based on time and human resources, and both are components that must and will increase the price. Obviously, when discussing sustainable luxury fashion, it is also of great value and importance that these processes take place under responsible social, economic, and environmental conditions.

Basically, one needs to recognize garments as luxury not purely because of the price or the sender in terms of brand value, but rather because of the story created in relation to a garment. It is about generating a humbleness and respect for the processes of making such products, and furthermore for the human beings who have created them.

As formerly mentioned, emotional products are actually able to affect one's behavior and even create change. In order to actually make a shift in producing and consuming mass to less and appreciating more *in* less, those specific stories—in which time and human presence constitutes the essence—need to be transparent and

---

[38]Thackara (2006).

touch the consumer on an emotional level. Naturally, a lot of luxury garments already implicitly fulfill or integrate themselves in these different suggestions, but that does not mean that the approaches should not be more prevalent when it comes to the understanding of the concept of luxury.

6.6   Luxury Consumer Versus Consuming Luxury

Consistent with reassessing the term of luxury, the consumer needs to be addressed as a 'person consuming luxury' instead of verbalizing him or her as a 'luxury consumer'. The concept of luxury consumers draws attention to an isolated group of people who can afford luxury goods and for whom these goods are mainly of a symbolic and status-related values, mediums used to express wealth. By inverting the word order as to convey these consumers are consuming luxury, the concept suddenly depicts a different openness toward a broader segment of people. From here on, it is necessary to educate the common consumer to think about quality and durability. It is also necessary to communicate and transparently convey the voyage of a luxury garment. Through this knowledge, the consumer can better be convinced about the long-term monetarily, as well as sustainable, advantages of buying fewer, but better garments, which can be expressed in numerous ways. The consumer simply needs to perceive garments as lifelong investments.

Changes in the approach to how one can articulate and assess luxury fashion and the consumption hereof are possible through knowledge and communication. The fashion industry needs to reengage with the customer on the basis of transparency and trust, so that he or she is not reduced to a mere consumer. According to Stuart Walker:

> Our lack of involvement in the design and making of objects, and our consequent gab in understanding, undoubtedly affect how we value them.[39]

Of course, it is naive to imagine a future scenario in which the consumer himself makes all garments; nevertheless, this is actually a growing tendency exemplified by the community of 'makers movement.' What is not a naïve though, is to strive for better-educated consumers that understand how things are made. Unquestionably this is an achievable goal—simply put it is necessary to:

> foster new relationships between the people who make things and the people who use them.[40]

As a result, these relationships might possibly lead to more emotionally engaged consumers, which again will lead to a more sustainable way of consuming.

---

[39]Walker (2006).
[40]Thackara (2006).

7 Conclusion

Creating sustainable luxury fashion is about touching the consumer, generating enduring emotional conditions, and accommodating and satisfying long-term needs. On this basis, it is necessary to create products, which in addition to possessing physical qualities and durable aesthetics are designed to emotionally engage the user with the product, ideally reaching the goal of irreplaceability. To simplify this concept, the consumer must be encouraged to emotionally engage when choosing a garment in order to successfully generate less replacement in a wardrobe.

The realization of such change demands a major initiative from both designers and the industry. Likewise, the consumer needs to evolve through a pleasure-based conviction that he or she is actively willing to contribute to such a development. Without the consumer's cooperation, this shift will simply remain an ideology.

The author would like to end this chapter by paraphrasing Kundera (2012) from the book *Immortality* and state that if one chooses to engage emotionally and committedly with one's garments, the state of immortality might be achieved, as a small part of one's soul is located within such belongings.

References

Cooper R, Sawaf A (1997) Executive E.Q. The Berkley Publishing Group, New York

Csikszentmihalyi M (2013) Creativity—the psychology of discovery and invention. Harper Perennial Modern Classics, New York

Den danske ordbog—moderne dansk sprog (2015) Det Danske Sprog- og Litteraturselskab. http://www.ordnet.dk. Assessed 3 May 2015

Fletcher K, Grose L (2012) Fashion and sustainability—design for change. Laurence King Publishing, London

Heskett J (2001) Past, present and future in design for industry. Des Issues 17(1):18–27

Heskett J (2002) Design—a very short introduction. Oxford University Press, New York

Holbech MS (2014) Vanessa Friedman: Bæredygtig mode giver ingen mening. www.fashionforum.dk. Accessed 3 May 2015

Kragelund M, Otto L (2005) Materialitet og Dannelse. Danish University of Education Press, Copenhagen

Kundera M (2012) Udødeligheden. Gyldendal, Copenhagen. Czech edition: Kundera M (1993) Nesmrtelnost (trans: Eva Andersen). Atlantis, Brno

Lawson B (2005) How designers think—The design process demystified. Architectural Press, Oxford

Norman DA (2005) Emotional design: why we love (or Hate) everyday things. Basic Books, New York

Olsen ET (2009) The Passage of time. In: Poidevin RL et al (eds) The routledge companion to metaphysics. Routledge, Abingdon, pp 440–448

Simon HA (1996) The sciences of the artificial. MIT Press, Cambridge

Svendsen LFH (2005) Mode—et filosofisk essay (trans: Wrang J). Klim, Aarhus. Norwegian edition: Svendsen LFH (2004) Mote—et filosofisk essay. Universitetsforlaget, Oslo

Thackara J (2006) In the bubble—designing in a complex world. MIT Press, Cambridge

Walker S (2006) Sustainable by design: explorations in theory and practice. Earthscan, London

The Devil Buys (Fake) Prada: Luxury Consumption on the Continuum Between Sustainability and Counterfeits

Nadine Hennigs, Christiane Klarmann and Franziska Labenz

Abstract Against the backdrop of the significant expansion in the luxury industry along with the ongoing process of the *luxurification of mass markets* and the *massification of luxury brands*, luxury brand managers act in the rising tension of satisfying the growing demand for luxury in the global marketplace and the effort to protect the uniqueness and exclusivity of their products. As a consequence, the alignment of luxury and sustainability is considered as a promising way to emphasize the key attributes of luxury such as heritage, timelessness, durability, and excellence in manufacturing and retailing. Nevertheless, in times of economic recession and widely available and often consumed counterfeit goods, the question arises whether the demand side is ready for the commitment to sustainability. In this context, the focus of our chapter is on the study of determinants of the "*dark side of luxury consumption,*" one of the largest challenges in luxury brand management: the increased demand for counterfeit branded products. The aim of the present study was to empirically investigate a multidimensional framework of counterfeit risk perception and counterfeit shopping behavior as perceived by distinct consumer segments. Even though price is often believed to be the main reason that causes counterfeit purchases, this study reveals that there are multifaceted reasons that affect consumer attitudes and behavior. Therefore, luxury brand managers have to respect and emphasize the deep-rooted values of the luxury concept: True luxury has to verify that it is more than shallow bling and superficial sparkle—the adoption of sustainability excellence is a promising strategy to demonstrate the credibility of luxury in offering superior performance in any perspective.

Keywords Sustainable luxury · Consumer complicity · Counterfeit consumption · Perceived risk

N. Hennigs (✉) · C. Klarmann · F. Labenz
Institute of Marketing and Management, Leibniz University of Hannover,
Koenigsworther Platz 1, 30167 Hannover, Germany
e-mail: hennigs@m2.uni-hannover.de

© Springer Science+Business Media Singapore 2016   99
M.A. Gardetti and S.S. Muthu (eds.), *Handbook of Sustainable Luxury
Textiles and Fashion*, Environmental Footprints and Eco-design
of Products and Processes, DOI 10.1007/978-981-287-742-0_6

1 Introduction

Against the backdrop of the significant expansion in the luxury industry along with the ongoing process of the *luxurification of mass markets* and the *massification of luxury brands*, luxury brand managers act in the rising tension of satisfying the growing demand for luxury in the global marketplace and the effort to protect the uniqueness and exclusivity of their products (Tynan et al. 2010; Hennigs et al. 2013a). The adoption of mass marketing strategies has led to intensified distribution and a downgrading strategy with an emphasis on mass production and vertical brand extensions related to accessories, perfume, and cosmetics (Dion and Arnould 2011). Delocalization to low-cost countries with poor labor standards and the availability of more accessible product lines based on licensing agreements have shifted the focus from deeper values to superficial logo conspicuousness (Kapferer and Michaut 2014). Facing the risks of brand dilution or overextension and the potential loss of brand equity through brand overexposure as well as the omnipresence of low-cost counterfeits and fake luxury products, the fundamental characteristics of the luxury concept are at stake (Hennigs et al. 2013a).

As a consequence, the alignment of luxury and sustainability is considered as a promising way to emphasize the key attributes of luxury such as heritage, timelessness, durability, and excellence in manufacturing and retailing (e.g., Bendell and Kleanthous 2007; Davies et al. 2012; Hennigs et al. 2013b; Janssen et al. 2013; Kapferer 2010). Nevertheless, in times of economic recession and widely available and often consumed counterfeit goods, the question arises whether the demand side is ready for the commitment to sustainability. For sure, following a sustainable strategy, luxury brands are expected to provide deeper value than a counterfeit luxury product which only tries to imitate the original. The essence of luxury is (or at least should be) far more than a nice and easy to copy logo. In the rising tension between the call for a more sustainable luxury on the one hand and an increasing demand for counterfeit goods on the other hand, the following questions arise:

> What does current consumption in the luxury market look like? Are luxury consumers willing to reconsider their emphasis on desirable logos? Are they prepared to become part of the sustainable luxury movement and adopt deeper values? What would be the necessary conditions for the co-existence of luxury and sustainability from the consumer perspective?

Reasoning this, the focus of our chapter is on the study of determinants of the *"dark side of luxury consumption,"* one of the largest challenges in luxury brand management: the increased demand for counterfeit branded products. Consumers often ignore the risks inherent in counterfeit activities and can be described as accomplices in crime, who actively seek counterfeit goods. The importance of focusing on the demand side becomes evident as all governmental actions to curtail counterfeiting will not be sufficient as long as counterfeiters face such an immense demand for their products. The aim of the present study was to begin filling this research gap by the following:

(a) examining factors that significantly influence counterfeit risk perception and counterfeit shopping behavior and

(b) identifying groups of consumers who differ in the specific reasons for acceptance of/resistance to counterfeit luxury goods.

The paper is structured as follows: The theoretical background for the concept of luxury and the phenomenon of counterfeiting will be provided in the next paragraph. Based on these insights and with reference to previous research on the demand side of counterfeiting luxury goods, the conceptual model and related hypotheses are presented. The methodology part outlines the instrument and sample used for the empirical study, before the results are presented and finally discussed regarding research and managerial implications as opportunities to develop strategies that aim to reduce the global appetite for counterfeits as important step on a promising way to align luxury (brands and consumers) with sustainability. In sum, the main results of this study underline that the shift of the luxury concept from exclusivity and exceptional craftsmanship to mass marketing strategies and superficial logo dominance is accompanied with less moral consumer concerns about counterfeit consumption. The more consumers attach deeper values to luxury goods, the less they are inclined to purchase counterfeits. Therefore, the awareness of a luxury brand's cultural heritage, the reassertion of the brand's own virtues, and the adoption of sustainability excellence are crucial elements of the credibility of the luxury concept.

2 Theoretical Background

Luxury goods as one of the fastest expanding product groups can traditionally be defined as *"goods for which the mere use or display of a particular branded product confers prestige on their owners, apart from any utility deriving from their function"* (Grossmann and Shapiro 1988, p. 82). Thus, for luxury brands it is essential to evoke exclusivity, brand identity, brand awareness as well as perceived quality from the consumer's perspective (Phau and Prendergast 2000). Therefore, luxury as a multidimensional construct is situational contingent and should follow an integrative understanding: *"What is luxury to one may just be ordinary to another"* (Phau and Prendergast 2000, p. 123). According to Vigneron and Johnson (1999), luxury brands are seen as the highest level of prestigious brands encompassing several physical and psychological values. In order to explain consumer behavior, the notion *"buying to impress others"* has long been a guiding principle for luxury brand managers. However, it has been found that in addition to interpersonal aspects such as snobbery (Leibenstein 1950; Mason 1992), personal factors such as hedonism and perfectionism (Dubois and Laurent 1994) as well as situational conditions (e.g., economic and societal factors) are particularly important. In addition to that, luxury consumers have become increasingly concerned about social and environmental issues (Cone 2009; Kleanthous 2011). In recent years, a paradigm shift has taken place in the luxury domain from "*conspicuous consumption*" to "*conscientious consumption*" (Cvijanovich 2011) leading to more critical and well-informed consumers (Sarasin 2012). The concept of luxury being traditionally based on high quality, superior durability, and deeper value is a perfect basis for the design and marketing of products that preserve fundamental social and environmental values (Kapferer 2010). Therefore, the concept of sustainability is of major importance for the management of luxury brands.

However, in contrast to the increasing consumers' awareness of social and environmental issues, a significant growth in the demand for counterfeit goods has to be considered in the luxury market as well. Counterfeits can be defined as "...*any manufacturing of a product which so closely imitates the appearance of the product of another to mislead a consumer that it is the product of another or deliberately offer a fake substitute to seek potential purchase from non-deceptive consumers*" (OECD 1998). The focus of this study is on non-deceptive counterfeiting which is prevailing in the luxury market (Nia and Zaichkowsky 2000) and stands for copies where consumers know or strongly suspect that the purchased product is not an original (Grossman and Shapiro 1988). Counterfeiting harms the legitimate producers and may result in a reduction of the exclusiveness of the genuine product which in turn could potentially erode consumers' confidence in a brand (Green and Smith 2002; Wilke and Zaichkowsky 1999). As a consequence, original brands face lost revenues and a loss of intangible values such as brand reputation and consumer goodwill (Bush et al. 1989). Nevertheless, most consumers disregard the negative effects counterfeiting entails (Phau et al. 2009a, b). Therefore, deepening the understanding of the customer perspective on counterfeit goods is central for the development of effective countermeasures because all actions to curtail counterfeit activities will not be sufficient as long as counterfeiters face such an immense demand for their products (Ang et al. 2001).

3 Conceptual Model and Related Hypotheses

In accordance with previous research dealing with the demand side of counterfeit goods (Ang et al. 2001; Huang et al. 2004), both psychological consumer traits and context-related aspects should be integrated into a single model. As illustrated in Fig. 1, the study presented here considers a combination of *personality factors* (i.e., variety seeking, personal integrity, moral judgment, and risk aversion) and *context-related factors* (i.e., luxury involvement, luxury value perception, and the trade-off between genuine and counterfeit luxury goods) as antecedents of consumers' *risk perception* toward counterfeits and actual *counterfeit shopping behavior*.

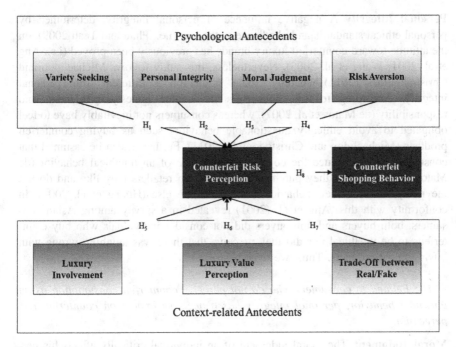

Fig. 1 The conceptual model

3.1 Psychological Antecedents

Variety Seeking In general, novelty seeking encompasses the desire of individuals to seek variety and difference (Phau and Teah 2009; Wang et al. 2005), whereas variety seeking in particular comprises the consumers demand for different things and a great deal of variety (Donthu and Gilliland 1996). Bringing variety into the context of luxury consumption, the well-documented luxury characteristics of rarity and exclusivity (Vigneron and Johnson 2004) may be connected to the consumer perceived variety. On the other hand, with reference to consumers who fear the hassle of being stuck with a "last-season" item (Wiedmann et al. 2007), luxury counterfeits as mass products which are often out of season would not be convenient to a high variety seeking consumer. Although various studies on counterfeit consumption exposed a negative influence of variety seeking on attitudes toward faked products (i.e., Wee et al. 1995), opposed to previous studies, this study conceives variety seeking as a desire for quality and less for quantity and therefore assumes a positive influence on the risk perception regarding counterfeits. It can be postulated that,

H_1: *Variety seeking in combination with the desire for exclusivity has a positive impact on counterfeit risk perception.*

Personal Integrity A negative influence of personal integrity, determined by personal ethical standards and obedience to the law (i.e., Phau and Teah 2009), on the attitude toward counterfeit luxury brands has previously been proved (i.e., Ang et al. 2001; Wang et al. 2005). Nevertheless, this study follows Michaelidou and Christodoulides (2011) whereby ethical obligation is different from personal integrity. According to this, consumers may value honesty, politeness, and responsibility (de Matos et al. 2007), whereas consumers not inevitably have to feel obligated to avoid ethically questionable behaviors such as buying counterfeit products (Michaelidou and Christodoulides 2011). Further, it can be assumed that consumers aim to reduce the cognitive dissonance of an unethical behavior (de Matos et al. 2007) or they purchase products from retailers they like and do not inevitably feel that their behavior harms someone else (Huang et al. 2004). In conformity with this, Ang et al. (2001) revealed in a survey among Asian consumers, both buyers and non-buyers did not consider individuals who buy counterfeits to be unethical nor did they perceive that there was anything wrong with buying faked products. Thus, we hypothesize,

H_2: *Related to consumers who do not perceive counterfeit consumption as an unethical behavior, personal integrity has a negative impact on counterfeit risk perception.*

Moral Judgment The moral judgment of an individual critically affects his perception as to why certain actions are perceived as morally just or preferred (Tan 2002). As the counterfeit supply side is often related to organized crime (Furnham and Valgeirsson 2007; Green and Smith 2002; Nill and Schultz 1996), consumer participation in a counterfeit transaction supports illegal activity (de Matos et al. 2007). According to this, it can be anticipated that consumers with a high standard of moral judgment may perceive a higher risk associated with counterfeit consumption, especially connected to individual and social issues. Accordingly,

H_3: *For consumers with high moral standards, moral judgment has a positive impact on counterfeit risk perception.*

Risk Aversion Considered as a personality variable and defined as the propensity to avoid taking risks (Zinkhan and Karande 1991), risk aversion can be seen as an important characteristic for discriminating between buyers and non-buyers of a product category (de Matos et al. 2007). Huang et al. (2004) already revealed a significant inverse relationship between risk averseness and attitude toward counterfeits. Focusing on counterfeit risk perception, it can be assumed that consumers with a high avoidance of taking risks perceive a significant higher financial, functional, social, and individual risk regarding faked products which presumably not offer the same value as the genuine version. Thus, it is suggested that,

H_4: *As the individual avoidance to take risks, risk aversion has a positive impact on counterfeit risk perception.*

3.2 Context-Related Antecedents

Luxury Involvement Understood as an internal state that indicates the amount of arousal, interest, or drive evoked by a particular stimulus or situation, involvement has been shown to influence purchasing behavior (Park and Mittal 1985). In terms of the average interest in a product category on a daily basis, a high level of product-class involvement leads to the consumer's willingness to spend more energy on consumption-related activities and hence make more rational decisions (Wilkie 1994; Zaichkowsky 1985). Therefore, high-involved consumers have a more favorable attitude to luxury goods in general and have stronger purchase intentions (Huang et al. 2004). Consequently, when they cannot afford the real item, consumers with a strong personal desire for luxury goods might be more likely to purchase the counterfeit alternative (Bloch et al. 1993; Phau and Teah 2009; Wilcox et al. 2008) and perceive a lower level of risk associated with this activity. It is expected that,

H_5: *Luxury involvement as the strong personal desire for luxury branded products has a negative impact on counterfeit risk perception.*

Luxury value perception With regard to consumption values that directly explain why consumers choose to either buy or avoid particular products (Sheth et al. 1991), different types of values influence consumers' purchase choices. In a luxury product context, the evaluation and propensity to purchase or consume luxury brands can be explained by the following four dimensions (Wiedmann et al. 2007, 2009):

- The financial dimension that addresses direct monetary aspects
- The functional dimension that refers to basic utilities as quality, uniqueness, and usability
- The individual dimension that addresses personal matters such as materialism, hedonism, and self-identity, and
- The social dimension that refers to aspects of status consumption and prestige orientation.

With reference to counterfeit luxury goods, it is expected that consumers who have a high value perception of genuine luxury goods are less willing to purchase counterfeits because it diminishes the idea that counterfeit consumption is a savvy shopper behavior and simultaneously enhances the perceived embarrassment potential (Wiedmann et al. 2012). Consequently, it can be assumed that the higher the consumer's value perception of the genuine luxury good, the more he or she is worried about the buying decision and has a higher risk perception of the counterfeit alternative. Reasoning this, it is hypothesized that,

H_6: *Luxury value perception related to the original product has a positive impact on counterfeit risk perception.*

Trade-off between Genuine and Counterfeit Luxury Goods Assuming that the market for counterfeit brands relies on consumers' desire for and evaluation of real luxury brands (Hoe et al. 2003; Penz and Stöttinger 2005), the individual choice decision between authentic and counterfeit products is influenced by a trade-off based on the combination of the price of the product (Furnham and Valgeirsson 2007), the perceived value of the product (Bloch et al. 1993; Furnham and Valgeirsson 2007), and the quality of the authentic product (Munshaw-Bajaj and Steel 2010). When presented with a choice between an authentic and a counterfeit luxury good, consumers who have a favorable opinion about the financial, functional, individual, and social value of the counterfeit alternative perceive purchasing counterfeits as an acceptable choice. Therefore, in the trade-off between authentic and counterfeit luxury products, it is expected that,

H_7: *For consumers who perceive counterfeits as an acceptable choice, the individual trade-off between real and fake luxury goods has a negative impact on counterfeit risk perception.*

3.3 Related Outcomes

Counterfeit Shopping Behavior With reference to the impact of consumers' counterfeit risk perceptions on actual counterfeit shopping behavior, literature suggests that consumers who perceive more risk in the counterfeit alternative are less likely to buy counterfeit goods (Albers-Miller 1999; Bloch et al. 1993; Nia and Zaichkowsky 2000). Understood as *"the consumer's perceptions of the uncertainty and adverse consequences of buying a product or service"* (Dowling and Staelin 1994, p. 119), consumers associate counterfeits with a higher level of risks that mediate consumers' evaluations of and feelings toward counterfeit purchases (Bamossy and Scammon 1985; Chakraborty et al. 1996). The perception of financial, functional, psychological, and social risks related to the purchase of a counterfeit will influence every stage of the consumer decision-making process (de Matos et al. 2007). Therefore,

H_8: *Counterfeit risk perception has a negative impact on actual and future counterfeit shopping behavior.*

4 Methodology

To measure the antecedents and behavioral outcomes of counterfeit risk perception in the context of our conceptual model, as shown in Table 1, we used existing and tested scales for assessing the *psychological antecedents* (i.e., variety seeking, personal integrity, moral judgment, risk aversion), the *context-related antecedents*

Table 1 The questionnaire scales

Scale	Author(s), year
Psychological antecedents	
Variety seeking	Donthu and Gilliland (1996)
Personal integrity	Ang et al. (2001)
Moral judgment	Tan (2002)
Risk aversion	Donthu and Gilliland (1996)
Context-related antecedents	
Luxury involvement	Beatty and Talpade (1994)
Luxury value perception	Sweeney and Soutar (2001), Wiedmann et al. (2009)
Trade-off between genuine and counterfeit good	In accordance to Wiedmann et al. (2009)
Related outcomes	
Counterfeit risk perception	Ang et al. (2001), Ha and Lennon (2006), Stone and Grønhaug (1993)
Counterfeit shopping behavior	Kressmann et al. (2003)

(i.e., luxury involvement, luxury value perception, trade-off between genuine and counterfeit good), and *related outcomes* (i.e., counterfeit risk perception, counterfeit shopping behavior). All items were rated on five-point Likert scales (1 = *strongly disagree* to 5 = *strongly agree*). The first version of our questionnaire dedicated to the investigation of the demand side of counterfeit luxury goods was face-validated using exploratory and expert interviews with six luxury researchers and six luxury consumers to check the length and layout of the questionnaire and the quality of the items used. To examine the research model based on the scales used in the questionnaire, personal interviews were considered most appropriate as data collection instrument for this study. To address the issue of social desirability bias and the respondent's inclination to conform to social norms, we preferred purposive sampling for which the units of observation are habitually luxury and/or counterfeit consumers. The recruitment of interviewees was organized by a personal invitation mail that was sent to members of a luxury consumer panel in Germany. Measured by market size, the German luxury market belongs to the top 3 global luxury markets (Roland Berger 2013). Germany is also of specific interest for the present study on counterfeiting, as German consumers have a particularly high exposure to counterfeit goods compared to other European countries: Following the Netherlands with Rotterdam as the biggest seaport in Europe, Germany, with the second (Hamburg) and fourth (Bremen) largest ports, detects the second largest volume of counterfeits entering the EU (European Commission 2009; UNODC 2010).

In the final sample, only those respondents were included who agreed to the statements that they are highly interested in the domain of luxury products and purchase luxury brands on a regular basis—either the original or the counterfeit alternative. Besides, all respondents in the final sample stated that they will purchase luxury brands again in the future. A total of 123 questionnaires were received

Table 2 Demographic profile of the sample

Variable		n	%
Age	18–25 years	86	71.7
	26–35 years	27	22.5
	36–55 years	6	5.0
	56–99 years	1	0.8
Gender	Male	46	37.4
	Female	74	60.2
Marital status	Single	108	87.8
	Married	11	8.9
	Widowed	1	0.8
Education	Lower secondary school	1	0.8
	Intermediate secondary school	6	4.9
	University entrance diploma	71	57.7
	University degree	42	34.1
Occupation	Full time	32	26.0
	Part time	5	4.1
	Pensioner and retiree	1	0.8
	Housewife and husband	2	1.6
	Job training	4	3.3
	Student	72	58.5
	Seeking work	3	2.4
Income	Very low income	3	2.4
	Low income	7	5.7
	Middle income	68	55.3
	High income	36	29.3
	Very high income	1	0.8

in January 2013; the sample characteristics are described in Table 2. Regarding gender distribution, 60.2 % of the respondents were female and 71.7 % of the participants were between 18 and 25 years of age, with 26.2 years as the mean age. The higher percentage of younger and female consumers may be attributed to the higher interest of female consumers in luxury brands and their willingness to participate in a study on luxury and counterfeit goods. Besides, due to budget restrictions and the question of affordability of genuine luxury, it can be assumed that this consumer group is more likely to choose the counterfeit alternative of a luxury good (Yoo and Lee 2009). With regard to educational level, 91.8 % of the sample had received a university entrance diploma or a university degree. With reference to the study context of luxury and counterfeit goods, 82.9 % of the respondents have already bought a genuine luxury product at least once and 56.9 % have already bought a counterfeit luxury product. Although this is not a representative one, with reference to the given research focus, the convenience sample used in this study offers a balanced set of data to empirically investigate consumer perceptions of counterfeit products.

5 Results and Discussion

SPSS 19.0 and SmartPLS 2.0 were used to analyze the data. In our exploratory study context of examining the drivers and outcomes of counterfeit risk perception, PLS path modeling was considered the appropriate method for the empirical tests of our hypotheses. With the primary objective of maximizing the explanation of the variance (or, equivalently, minimizing the error) in the dependent constructs of a structural equation model (Henseler et al. 2009), PLS integrates principal component analysis with multiple regression (Hahn et al. 2002).

To assess common method variance, following Podsakoff et al. (2003), we used Harman's (1976) one-factor test to determine whether a single factor accounted for most of the covariance in the relationships between the independent and dependent variables. A principal component factor analysis with varimax rotation revealed a 9-factor structure with no general factor present (the first factor accounted for 9.5 % of the variance). Thus, no single factor accounted for a majority of the covariance in the variables, so the common method variance was unlikely to present a significant problem in our study. The results of the measurement of the constructs, the test of our hypotheses, and the cluster segments are described below.

Measurement of Constructs For a reliable and valid measurement of the latent variables, we followed the suggestions of China (1998). For all factors, our results show sufficiently high factor loadings. Additionally, the average variance extracted (AVE), the reliability tests (Cronbach's alpha, indicator reliability, factor reliability), and the discriminant validity (Fornell–Larcker criterion) revealed satisfactory results (see Table 3).

Evaluation of Structural Relations To test our hypotheses, we conducted a PLS path modeling analysis with casewise replacement and a bootstrapping procedure (individual sign changes; 123 cases and 1000 samples). As illustrated in Fig. 2 and Table 4, the assessment of the aggregate PLS path coefficients in the inner model results in statistically significant relations ($p < 0.01$). Referring to *psychological antecedents*, the latent variables *Variety Seeking, Moral Judgment,* and *Risk Aversion* reveal a positive and significant relationship to the latent variable *Counterfeit Risk Perception,* providing full support for hypotheses H_1, H_3, and H_4. As suggested, the impact of *Personal Integrity* on *Counterfeit Risk Perception* was significant and negative, and this is supportive of H_2. With reference to the *context-related antecedents*, in hypothesis H_5, we postulated that *Luxury Involvement* has a negative impact on *Counterfeit Risk Perception*. The results reveal full support for H_5; the effects between *Luxury Involvement* and *Counterfeit Risk Perception* are significant and negative. Regarding H_6, as suggested, there is a significantly positive impact of *Luxury Value Perception* on *Counterfeit Risk Perception*. Furthermore, supportive of H_7, the results show a significant and negative relation between the *Trade-Off between Real and Fake* and *Counterfeit Risk Perception*. Consumers who made their choice in favor of counterfeit goods perceive such purchases as less risky. Besides, the assessment of the impact of *Counterfeit Risk*

Table 3 Evaluation of the measurement models

Factor	Cronbach's alpha	Factor loadings	t-Value	Composite reliability	AVE	Fornell–Larcker criterion
Psychological antecedents						
F1 variety seeking						
I like to try different things	0.873	0.820	28.981	0.912	0.784	0.784 > 0.132
I like a great deal of variety		0.913	69.346			
I like new and different styles		0.919	58.139			
F2 personal integrity						
I consider honesty an important human trait	0.766	0.868	18.229	0.863	0.678	0.678 > 0.132
I consider politeness an important human trait		0.827	12.130			
I consider responsibility an important human trait		0.773	7.896			
F3 moral judgment						
In my opinion, it is morally wrong to buy a counterfeit instead of the genuine product	0.907	0.865	287.312	0.940	0.840	0.840 > 0.309
It is morally wrong to buy counterfeit luxury goods		0.894	401.752			
There are ethical reasons against buying counterfeit luxury products		0.822	54.023			
F4 risk aversion						
I would rather be safe than sorry	0.715	0.796	30.786	0.839	0.638	0.638 > 0.074
I want to be sure before I purchase anything		0.680	20.542			
I avoid risky things		0.905	74.405			

(continued)

Table 3 (continued)

Factor	Cronbach's alpha	Factor loadings	t-Value	Composite reliability	AVE	Fornell–Larcker criterion
Context-related antecedents						
F5 luxury involvement						
I am very interested in luxury goods	0.731	0.913	71.115	0.846	0.658	0.658 > 0.323
Luxury goods play an important role in my life		0.925	47.251			
I never get bored when people talk about luxury goods		0.535	6.499			
F6 luxury value perception						
The price of a luxury good matches its quality	0.694	0.512	12.655	0.780	0.291	0.291 > 0.271
Luxury products are made of high quality		0.640	20.875			
A luxury good satisfies my needs		0.481	12.521			
A luxury product cannot be sold in supermarkets		0.605	19.867			
The luxury brands I buy must match what and who I really am		0.309	6.567			
For me luxury goods are truly delightful		0.653	24.661			
I like a lot of luxury in my life		0.514	13.956			
I like to know what brands and products make a good impression on others		0.400	7.888			
Luxury goods help to make a good impression on others		0.638	21.851			

(continued)

Table 3 (continued)

Factor	Cronbach's alpha	Factor loadings	t-Value	Composite reliability	AVE	Fornell–Larcker criterion
F7 Trade-off between genuine and counterfeit good referring to...						
Functionality	0.834	0.459	11.938	0.866	0.336	0.336 > 0.271
Quality		0.643	27.799			
Usability		0.572	20.710			
Uniqueness		0.595	27.260			
Prestige		0.546	21.710			
My self-concept		0.690	43.364			
Personal gratification		0.667	33.584			
Visual attributes: Logo and Brand Insignia		0.463	12.958			
Conspicuousness		0.463	11.557			
Social status		0.639	28.104			
Self-realization		0.687	36.621			
Belonging to friends		0.535	19.704			
Ethical aspects		0.506	19.710			
Related Outcomes						
F8 Counterfeit risk perception						
If I bought a counterfeit luxury product, I would be concerned that I really would not get my money's worth from this product	0.806	0.727	41.250	0.866	0.565	0.565 > 0.401
The quality of a fake product will be very poor		0.666	32.337			
I would not feel very comfortable wearing a fake product in public		0.831	79.758			
People in my social environment do not appreciate counterfeit luxury goods		0.759	40.026			
All in all, I consider buying a counterfeit luxury product as very risky		0.767	60.991			

(continued)

Table 3 (continued)

Factor	Cronbach's alpha	Factor loadings	t-Value	Composite reliability	AVE	Fornell–Larcker criterion
F9 Counterfeit shopping behavior						
I have already bought counterfeit luxury products	0.468	0.529	10.803	0.700	0.371	0.371 > 0.332
I have bought counterfeit luxury products several times		0.582	14.426			
I consider buying counterfeit luxury goods in the future		0.730	36.228			
I do not intend to buy genuine luxury goods in the future		0.578	14.235			

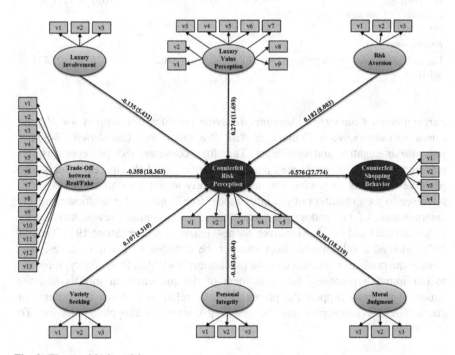

Fig. 2 The empirical model

Table 4 Evaluation of the structural relations

Exogenous LV → Endogenous LV	Original sample	Sample mean	Standard deviation	Standard error	T statistics
Psychological antecedents					
H_1: Variety seeking → Counterfeit risk perception	0.107	0.102	0.025	0.025	4.310
H_2: Personal integrity → Counterfeit risk perception	−0.163	−0.154	0.025	0.025	6.494
H_3: Moral judgment → Counterfeit risk perception	0.383	0.381	0.021	0.021	18.219
H_4: Risk aversion → Counterfeit risk perception	0.182	0.181	0.023	0.023	8.003
Context-related antecedents					
H_5: Luxury involvement → Counterfeit risk perception	−0.135	−0.133	0.025	0.025	5.432
H_6: Luxury value perception → Counterfeit risk perception	0.274	0.273	0.023	0.023	11.693
H_7: Trade-off real/fake → Counterfeit risk perception	−0.358	−0.361	0.020	0.020	18.363
Related outcomes					
H_8: Risk perception → Shopping behavior	−0.576	−0.576	0.021	0.021	27.774

Perception on *Counterfeit Shopping Behavior* provides full support for H_8; the causal relation between *Counterfeit Risk Perception* and *Counterfeit Shopping Behavior* is negative and significant. Therefore, consumer risk perception is significant in influencing counterfeit purchase intention and behavior; consumers who perceive more risk in counterfeits are less likely to purchase these goods. With reference to the evaluation of the inner model (see Table 5), the coefficients of the determination of the endogenous latent variables (R-square) reveal satisfactory values at 0.603 and 0.332. Moreover, Stone–Geisser Q-square (Stone 1974; Geisser 1974) yielded a value higher than zero for the endogenous latent variables, suggesting the predictive relevance of the explanatory variables. In summary, referring to our initial hypotheses, the assessment of the measurement models and the structural relations support the proposed causal relations between antecedents of counterfeit risk perception and the resulting counterfeit shopping behavior. To

Table 5 Evaluation of the inner model

Endogenous LV	R^2	Q^2
Risk perception	0.603	0.026
Counterfeit shopping behavior	0.332	0.113

develop appropriate strategies aimed at different types of genuine and counterfeit luxury consumers, in a next step, we used cluster analysis in conjunction with discriminant analysis.

Types of Genuine and Counterfeit Luxury Consumers To conduct the cluster analysis, the factor scores for each respondent were saved. In our analysis, we used a combination of Ward's method of minimum variance and non-hierarchical k-means clustering. The results strongly suggested the presence of four clusters. With regard to classification accuracy, we also used discriminant analysis to check the cluster groupings once the clusters were identified; 94.3 % of the cases were assigned to their correct groups, validating the results of cluster analysis for the useful classification of consumer subgroups based on the factors included in the model. To develop a profile of each market segment, more detailed information was obtained by examining the factor scores cross-tabulated by cluster segment, as presented in Table 6. Based on the variables from which they were derived, the four clusters were labeled as follows:

Cluster 1: The Luxury Lovers with a mean age of 27.5 years form 13.9 % of the sample, with 17.6 % male and 82.4 % female respondents and the highest income level of all groups. Referring to our study context, 88.2 % state that they purchase genuine luxury goods on a regular basis; 23.5 % have already bought a counterfeit luxury product—this is the smallest percentage of all groups. When presented with a choice of a genuine or a counterfeit luxury product, all respondents in this group prefer the authentic alternative. Regarding future behavior, 82.4 % intend to buy genuine luxury goods and 100 % refrain from buying counterfeit products. Taken as a whole, 82.4 % state, *"All in all, I consider buying a counterfeit luxury product as very risky."* Typical consumers in this cluster can be considered as non-consumers

Table 6 Cluster Means

	Factor means cluster 1	Factor means cluster 2	Factor means cluster 3	Factor means cluster 4	F	Sig.
F1 variety seeking	0.492	−0.176	−0.109	0.171	2.474	0.065
F2 personal integrity	0.400	−0.362	0.132	0.234	5.042	0.003
F3 moral judgment	0.218	−0.642	0.377	−0.368	27.008	0.000
F4 risk aversion	0.672	−0.490	−0.100	0.335	7.887	0.000
F5 luxury involvement	0.988	−0.150	−0.810	0.517	29.141	0.000
F6 luxury value perception	1.258	−0.334	−0.742	0.607	43.443	0.000
F7 trade-off	−1.313	0.556	0.261	−0.249	23.058	0.000
F8 counterfeit risk perception	1.605	−0.803	0.059	−0.024	48.223	0.000
F9 counterfeit shopping behavior	−0.883	1.007	−0.541	−0.063	39.347	0.000

of fake luxuries, as evidenced by the highest ratings for both psychological and context-related drivers of counterfeit risk perception and shopping behavior. Significantly more than others, they value the characteristics of authentic luxury and are not likely to take the risks associated with counterfeits.

Cluster 2: The Counterfeit Accomplices with a mean age of 26.7 years form 29.5 % of the sample, with 47.2 % male and 52.8 % female respondents and the lowest income level of all groups. Overall, 86.1 % of the respondents in this group state that they have already bought a genuine luxury product at least once, and, with the highest percentage of all groups, 86.1 % have already bought a counterfeit luxury good. In the trade-off between genuine and counterfeit luxury, consumers in this group are merely undecided or choose the counterfeit product (62.9 %). As evidenced by lowest factor mean scores on counterfeit risk perception, they do not perceive counterfeit shopping as being very risky (91.7 %). Referring to their buying intentions and related to the highest mean scores for counterfeit shopping behavior, 63.9 % intend to buy authentic luxury goods and 69.4 % consider buying a fake alternative.

Cluster 3: The Inexperienced Moralists with a mean age of 26.8 years comprise 30.3 % of the sample, with 40.5 % male and 59.5 % female respondents and middle income. In sum, 75.7 % of these respondents state that they possess genuine luxury goods and 35.1 % have already bought a counterfeit luxury product. As indicated by lowest mean scores for luxury involvement and luxury value perception, when they have to choose between genuine and counterfeit luxury products, only 54.1 % prefer the authentic product. Even though mean scores for moral judgment and counterfeit risk perception are second highest of all groups, they do not perceive shopping for counterfeits as being very risky (70.3 %). In the future, 81.1 % consumers of this group intend to buy genuine luxury goods, whereas 13.5 % consider buying counterfeits as a possible alternative.

Cluster 4: The Value-conscious Waverer with a mean age of 24.3 years comprises 26.2 % of the sample, with 40.6 % male and 59.4 % female respondents with middle to high income. In this cluster, as evidenced by second highest ratings for luxury value perception and luxury involvement, 87.5 % state that they possess genuine luxury goods and 65.6 % have already bought a counterfeit luxury product. In the trade-off between real and fake, 78.1 % choose the genuine luxury product over the counterfeit alternative—however, only 21.9 % perceive the purchase of a counterfeit as very risky. Referring to future behavior, 81.3 % prefer buying genuine luxuries and 18.8 % intend to buy counterfeit luxury goods.

With reference to our results and due to the fact that, in accordance with existing research (e.g., Furnham and Valgeirsson 2007), the individual perception of counterfeits was shown to be more important for consumer behavior than ethical or legal considerations, we hope that this study is another motivational basis for ongoing research in the area of consumer perception and behavior toward genuine and counterfeit luxury goods as outlined in the following section.

6 Conclusion

The global impact of counterfeiting is increasing at an alarming rate; its effects are perceptible at both macro- and microeconomic level. Governments, supranational organizations, and industry associations have undertaken considerable efforts to curtail the illegitimate business through IPR protection and law enforcement. Nevertheless, an attempt where countermeasures focus on the supply side only falls short; any remedy will be insufficient as long as there is a maintained demand for counterfeit products. A better understanding of the specific consumer motivation for purchasing these goods builds the basis for the development of strategies that aim to reduce the global appetite for counterfeits.

The primary goal of this chapter was to explore a multidimensional framework of counterfeit risk perception and counterfeit shopping behavior as perceived by distinct consumer segments. Even though price is often believed to be the main reason that causes counterfeit purchases, this study reveals that there are multi-faceted reasons that affect consumer attitudes and behavior. In this context, the results indicate that counterfeit risk perception negatively and significantly affects counterfeit shopping behavior. Moreover, the results reveal that the antecedents of counterfeit risk perception can be divided into two groups: psychological ante-cedents as a combination of personality factors and antecedents related to the context of genuine and counterfeit luxury goods. It has to be noted that the sample used in this study is not a representative one, and due to the limited generalizability of the results, it is reasonable to replicate the study with a large sample of typical luxury (counterfeit) consumers in different luxury markets to gain more differen-tiated results. However, the results of this study reveal interesting insights that lead to the following implications for research and management in the luxury industry.

From a managerial perspective, our study may form an appropriate basis to develop distinct strategies that aim to reduce the global appetite for counterfeits addressing cluster-specific differences between *luxury lovers, counterfeit accom-plices, inexperienced moralists,* and *value-conscious waverers.* The results indicate that countermeasures focusing on the price only fall short because counterfeit attitudes and consumption are driven by an individual combination of psychological and context-related antecedents. Therefore, the key challenge is to identify and address the specific risks and responsibilities associated with counterfeit con-sumption by distinct consumer groups, raise ethical considerations, display the negative consequences for society, and convince them that—compared to the value of owning genuine luxury (i.e., *"the taste and face of having the original,"* Gentry et al. 2006)—on the long run, counterfeit products are not worth the money. Nevertheless, the luxury industry has to take the lead and prove that the value of luxury is more than the use of logos. As long as luxury brand managers ignore the core principles of luxury such as exclusivity, superior craftsmanship, and excep-tional quality, but continue to follow trading down strategies in the pursuit for market growth and massive profits, *"consumers lose trust and respect in the brand and thus feel little guilt over counterfeit purchases"* (Kapferer and Michaut 2014,

p. 62). As a consequence, consumers with a desire for a prominent logo tag often choose the counterfeit that is in their opinion almost indistinguishable from the original, but much cheaper. Therefore, luxury brand managers have to respect and emphasize the deep-rooted values of the luxury concept such as tradition, heritage, exceptional quality, and uniqueness. True luxury has to verify that it is more than shallow bling and superficial sparkle—the adoption of sustainability excellence is a promising strategy to demonstrate the credibility of luxury in offering superior performance in any perspective.

References

Albers-Miller ND (1999) Consumer misbehaviour: why people buy illicit goods. J Consum Mark 16(3):273–287

Ang SH, Cheng PS, Lim EAC, Tambyah SK (2001) Spot the difference: consumer responses towards counterfeits. J Consum Mark 18(3):219–235

Bamossy G, Scammon DL (1985) Product counterfeiting: consumers and manufacturers beware. Adv Consum Res 12(1):334–340

Beatty SE, Talpade S (1994) Adolescent influence in family decision making: a replication with extension. J Consum Res 21(2):332–341

Bendell J, Kleanthous A (2007) Deeper luxury: quality and style when the world matters. Available via WWF. http://www.wwf.org.uk/deeperluxury. Accessed 05 Feb 2015

Bloch PH, Bush RF, Campbell L (1993) Consumer "accomplices" in product counterfeiting: a demand side investigation. J Consum Mark 10(4):27–36

Bush RF, Bloch PH, Dawson S (1989) Remedies for product counterfeiting. Bus Horiz 32(1): 59–65

Chakraborty G, Allred AT, Bristol T (1996) Exploring consumers' evaluations of counterfeits: the roles of country of origin and ethnocentrism. Adv Consum Res 23:379–384

Chin WW (1998) The partial least squares approach to structural equation modeling. In: Mar-coulides G (ed) Modern methods for business research. Mahwah, NJ, pp 295–358

Commission European (2009) Report on EU customs enforcement of intellectual property rights, results at the European border—2008. European Commission, Brussels

Cone (2009) Consumer environmental survey fact sheet. http://www.coneinc.com/stuff/contentmgr/files/0/56cf70324c53123abf75a14084bc0b5e/files/2009_cone_consumer_environmental_survey_release_and_fact_sheet.pdf. Accessed 25 Jan 2015

Cvijanovich M (2011) Sustainable luxury: oxymoron? Lecture in luxury and sustainability. http://www.mcmdesignstudio.ch/files/Guest%20professor%20Lucern%20School%20of%20Art%20%20and%20Design.pdf. Accessed 25 January 2015

Davies IA, Lee Z, Ahonkhai I (2012) Do consumers care about ethical luxury? J Bus Ethics 106 (1):37–51

de Matos C, Augusto C, Ituassu T, Rossi CAV (2007) Consumer attitudes toward counterfeits: a review and extension. J Consum Mark 24(1):36–47

Dion D, Arnould EJ (2011) Retail luxury strategy: assembling charisma through art and magic. J Retail 87(4):502–520

Donthu N, Gilliland D (1996) Observations: the infomercial shopper. J Advertising Res 36:69–76

Dowling GR, Staelin R (1994) Model of perceived risk and intended risk-handling activity. J Consum Res 21(1):119–134

Dubois B, Laurent G (1994) Attitudes toward the concept of luxury: an exploratory analysis. Asia Pac Adv Consum Res 1:273–278

Furnham A, Valgeirsson H (2007) The effect of life values and materialism on buying counterfeit products. J Socio Econ 36:677–685

Geisser S (1974) A predictive approach to the random effect model. Biometrika 6(1):101–107

Gentry JW, Putrevu S, Shultz CJ (2006) The effects of counterfeiting on consumer search. J Consum Behav 5:1–12

Green RT, Smith T (2002) Countering brand counterfeiters. J Int Mark 10(4):89–106

Grossman GM, Shapiro C (1988) Foreign counterfeiting of status goods. Q J Econ 103(1):79–100

Ha S, Lennon SJ (2006) Purchase intent for fashion counterfeit products: ethical ideologies, ethical judgments, and perceived risks. Clothing Text Res J 24(4):297–315

Hahn C, Johnson MD, Herrmann A, Huber F (2002) Capturing customer heterogeneity using a finite mixture PLS approach. Schmalenbach Bus Rev 54(3):243–269

Harman HH (1976) Modern factor analysis. 3rd edn. University of Chicago Press, Chicago

Hennigs N, Wiedmann K-P, Behrens S, Klarmann C (2013a) Unleashing the power of luxury: antecedents of luxury brand perception and effects on luxury brand strength. J Brand Manage 20(8):705–715

Hennigs N, Wiedmann K-P, Klarmann C, Behrens S (2013b) Sustainability as part of the luxury essence: delivering value through social and environmental excellence. J Corp Citiznsh 52 (11):25–35

Henseler J, Ringle CM, Sinkovics RR (2009) The use of partial least squares path modeling in international marketing. Adv Int Mark 20:277–319

Hoe L, Hogg G, Hart S (2003) Faking it: counterfeiting and consumer contradictions. Eur Adv Consum Res 6:60–67

Huang JH, Lee BCY, Hoe SH (2004) Consumer attitude toward gray market goods. Int Mark Rev 21(6):598–614

Janssen C, Vanhamme J, Lindgreen A, Lefebvre C (2013) The catch-22 of responsible luxury: effects of luxury product characteristics on consumers' perception of fit with corporate social responsibility. J Bus Ethics. doi:10.1007/s10551-013-1621-6

Kapferer J-N (2010) All that glitters is not green: the challenge of sustainable luxury. The European Business Review November/December

Kapferer J-N, Michaut A (2014) Luxury counterfeit purchasing: the collateral effect of luxury brands' trading down policy. J Brand Strategy 3(1):59–70

Kleanthous A (2011) Simply the best is no longer simple. The raconteur—sustainable luxury, July 2013

Kressmann F, Herrmann A, Huber F, Magin S (2003) Dimensionen der Markeneinstellung und ihre Wirkung auf die Kaufabsicht. Die Betriebswirtschaft 63(4):401–418

Leibenstein H (1950) Bandwagon, snob, and Veblen effects in the theory of consumers' demand. Quart J Econ 64:183–207

Mason RS (1992) Modeling the demand for status goods. Working paper, Department of business and management studies, University of Salford, St Martin's, New York

Michaelidou N, Christodoulides G (2011) Antecedents of attitude and intention towards counterfeit. J Mark Manage 27(9–10):976–991

Munshaw-Bajaj N, Steel M (2010) Exploring consumer choices in shopping for authentic and counterfeit goods. http://www.anzmac.org/conference_archive/2010/pdf/anzmac10Final00149. pdf. Accessed 23 April 2013

Nia A, Zaichkowsky JL (2000) Do counterfeits devalue the ownership of luxury brands? J Prod Brand Manage 9(7):485–497

Nill A, Shultz CJ (1996) The scourge of global counterfeiting. Bus Horiz 39(6):37–42

OECD (1998) The economic impact of counterfeiting. http://www.oecd.org/dataoecd/11/11/ 2090589.pdf. Accessed on 23 Jan 2015

Park CW, Mittal B (1985) A theory of involvement in consumer behavior: problems and issues. Res Consum Behav 1:201–231

Penz E, Stöttinger B (2005) Forget the real thing-take the copy! An explanatory model for the volitional purchase of counterfeit products. Adv Consum Res 32:568–575

Phau I, Prendergast G (2000) Consuming luxury brands: the relevance of the rarity principle. J Brand Manage 8:122–138

Phau I, Sequeira M, Dix S (2009a) Consumers' willingness to knowingly purchase counterfeit products. Direct Mark Int J 3(4):262–281

Phau I, Sequeira M, Dix S (2009b) To buy or not to buy a "counterfeit" Ralph Lauren polo shirt: the role of lawfulness and legality toward purchasing counterfeits. Asia Pac J Bus Adm 1(1):68–80

Phau I, Teah M (2009) Devil wears (counterfeit) Prada: a study of antecedents and outcomes of attitudes towards counterfeits of luxury brands. J Consum Mark 26(1):15–27

Podsakoff PM, MacKenzie SM, Lee J, Podsakoff NP (2003) Common method variance in behavioral research—a critical review of the literature and recommended remedies. J Appl Psychol 88:879–903

Roland Berger (2013) Meisterkreis index 2013 available via Roland Berger. http://www. rolandberger.com/media/pdf/Roland_Berger_Meisterkreis_Luxus_Index_20130225.pdf. Accessed 23 Apr 2013

Sarasin (2012) The quest for authenticity—can luxury brands justify a premium price? Available via Basel: Bank Sarasin & Co. http://www.sarasin.com. Accessed 15 Jan 2014

Sheth JN, Newman BI, Gross BL (1991) Why we buy what we buy: a theory of consumption values. J Bus Res 22:159–170

Stone M (1974) Cross-validatory choice and assessment of statistical predictions. J Roy Stat Soc 36(2):111–147

Stone RN, Grønhaug K (1993) Perceived risk: further considerations for the marketing discipline. Eur J Mark 27(3):39–50

Sweeney JC, Soutar GN (2001) Consumer perceived value: the development of a multiple item scale. J Retail 77(2):203–220

Tan B (2002) Understanding consumer ethical decision making with respect to purchase of pirated software. J Consum Mark 19(2):96–111

Tynan C, McKechnie S, Chhuon C (2010) Co-creating value for luxury brands. J Bus Res 63 (11):1156–1163

United Nations Office on Drugs and Crime, UNODC (2010) The globalization of crime: a transnational organized crime threat assessment. UNODC, Vienna

Vigneron F, Johnson LW (1999) A review and a conceptual framework of prestige-seeking consumer behaviour. Acad Mark Sci Rev 1:1–15

Vigneron F, Johnson LW (2004) Measuring perceptions of brand luxury. J Brand Manage 11 (6):484–506

Wang F, Zhang H, Zang H, Ouyang M (2005) Purchasing pirated software: an initial examination of Chinese consumers. J Consum Mark 22(6):340–351

Wee CH, Ta SJ, Cheok KH (1995) Non-price determinants of intention to purchase counterfeit goods: an exploratory study. Int Mark Rev 12(6):19–46

Wiedmann K-P, Hennigs N, Siebels A (2007) Measuring consumers' luxury value perception: a cross-cultural framework. Acad Mark Sci Rev 7:1–21

Wiedmann K-P, Hennigs N, Siebels A (2009) Value-based segmentation of luxury consumption behavior. Psychol Mark 26(7):625–651

Wiedmann K-P, Hennigs N, Klarmann C (2012) Luxury consumption in the trade-off between genuine and counterfeit goods: what are the consumers' underlying motives and value-based drivers? J Brand Manage 19:544–566

Wilcox K, Kim HM, Sen S (2008) Why do consumers buy counterfeit luxury brands? J Mark Res 46(2):247–259

Wilke R, Zaichkowsky JL (1999) Brand imitation and its effects on innovation, competition, and brand equity. Bus Horiz 42(6):9–18

Wilkie WL (1994) Consumer behavior, 3rd edn. Wiley, New York

Yoo B, Lee SH (2009) Buy genuine luxury fashion products or counterfeits? Adv Consum Res 36:280–286

Zaichkowsky JL (1985) Measuring the involvement construct. J Consum Res 12(3):341–352

Zinkhan GM, Karande KW (1991) Cultural and gender differences in risk-taking behavior among american and spanish decision makers. J Social Psychol 131(5):741–742

The Luxury of Sustainability: Examining Value-Based Drivers of Fair Trade Consumption

Steffen Schmidt, Nadine Hennigs, Stefan Behrens and Evmorfia Karampournioti

Abstract Green consumption has evolved into consumption that also addresses ethical factors. The twenty-first century is perceived to reflect the emancipation of the ethical consumer, who is *"shopping for a better world"* (Low and Davenport in J Consum Behav 6(5):336–348, 2007). The rising consumer demand for ethical alternatives is present in all product categories, and—reasoning that the concept of sustainability with aspects such as exclusivity and rareness shares similar values with the concept of luxury—the aim of our study is to examine the luxury of sustainability against the backdrop of the research questions concerning a proposed similarity of consumer associations between luxury and ethical products. As specific context, we have chosen the orientation to and acceptance of Fair Trade products. In detail, the present study empirically investigates a multidimensional framework of intrapersonal Fair Trade orientation, fair-trade-oriented luxury perception, and fair-trade-oriented customer perceived value with reference to the recommendation of Fair Trade products. The first contribution of our research is to provide a conceptual framework of value-based drivers of Fair Trade product perception against the backdrop of the luxury concept. Second, the empirical findings of the applied partial least squares equation modeling (PLS-SEM) contribute to the understanding of consequences of Fair-Trade-based perception. Customers who reveal a high luxury perception of Fair Trade products are strengthened in the Fair Trade idea which results in higher customer perceived value as well as the willingness to recommend Fair Trade products. Third, the data analysis of the applied PLS-SEM approach demonstrates that positive Fair Trade behavior is influenced by direct and indirect effects. With reference to the conducted study, being an active promoter of Fair Trade products is directly determined by the customers' perceived product value (benefit in relation to cost), but also directly and indirectly affected by the overall luxury product perception as well as the intrapersonal level of Fair Trade orientation.

Keywords Sustainable luxury · Fair Trade · Customer perceived value

S. Schmidt (✉) · N. Hennigs · S. Behrens · E. Karampournioti
Institute of Marketing and Management, Leibniz University of Hannover,
Koenigsworther Platz 1, 30167, Hannover, Germany
e-mail: schmidt@m2.uni-hannover.de

© Springer Science+Business Media Singapore 2016   121
M.A. Gardetti and S.S. Muthu (eds.), *Handbook of Sustainable Luxury Textiles and Fashion*, Environmental Footprints and Eco-design of Products and Processes, DOI 10.1007/978-981-287-742-0_7

1   Introduction

Green consumption and the avoidance of products that are likely to "*endanger the health of the consumer or other; cause significant damage to the environment during manufacture, use or disposal; consume a disproportionate amount of energy; cause unnecessary waste; use materials derived from threatened species or environments; involve unnecessary use—or cruelty to animals; adversely affect other countries*" (Elkington and Hailes 1989, p. 5) have evolved into consumption that also addresses ethical factors (Strong 1996). The twenty-first century is perceived to reflect the emancipation of the ethical consumer (e.g., Nicholls 2002), who is "*shopping for a better world*" (Low and Davenport 2007, p. 336). That said, consumers show a higher level of sustainability awareness. Indeed, sustainability is a worldwide social movement that has its origin in the late 1970s (Peet and Watts 2002), which is closely related to the global solidarity movement and corporate responsibility movement (Kates et al. 2005). It can be stated that the ethical-oriented buyer demands that products are not only friendly to the environment but also friendly to the people who produce them (Rosenbaum 1993). Sustainability awareness includes ethical issues such as environmental, animal welfare and societal concerns.

As a result, marketing managers are beginning to realize the importance of customer ethics and values and how meeting ethical demands is critical if they wish to gain a competitive advantage (Browne et al. 2000). However, when dealing with ethical purchases, results reveal a substantial gap between consumer attitudes, buying intentions, and effective behavior (e.g., De Pelsmacker et al. 2005a; King and Bruner 2000; Strong 1996). Even if studies suggest that consumers prefer to buy ethical products (e.g., Creyer and Ross 1997; Mohr et al. 2001), only a niche actually buys them (Cowe and Williams 2000). Therefore, the question remains of how to meet the consumers' expectations toward sustainable and green products and realize the transfer of positive perception to actual buying behavior. Consequently, a more thorough understanding of consumer ethical orientation and related links to ethical buying behavior is necessary (Papaoikonomou et al. 2011).

The rising consumer demand for ethical alternatives is present in all product categories, and—reasoning that the concept of sustainability with aspects such as exclusivity and rarity shares similar values with the concept of luxury (Phau and Prendergast 2000; Janssen et al. 2013)—the aim of our study is to examine the luxury of sustainability along the following questions: *Are there co-existing consumer associations with regard to luxury and ethical products? Can the sustainability sector learn from the management of luxury brands?* As a specific context, we have chosen the orientation to and acceptance of Fair Trade products. In the area of ethical consumerism and among the different social, environmental, and organic labels, Fair Trade has been the fastest growing sector, with international sales that increase by more than 15 % per year (FLO 2010).

Against the backdrop of the aim of the present work, a multidimensional framework of intrapersonal Fair Trade orientation, fair-trade-oriented luxury perception, and fair-trade-oriented customer perceived value with reference to the recommendation of Fair Trade products is conceptualized. In detail, partial least squares equation modeling (PLS-SEM) is applied to assess the introduced conceptual model empirically in order to contribute to the understanding of consequences of fair-trade-based perception.

The paper is structured as follows: First, a conceptual framework is derived and provided in the next paragraph. Particularly, the theoretical background of all core elements of the introduced model, namely fair-trade-oriented luxury perception, intrapersonal Fair Trade orientation, fair-trade-oriented customer perceived value, and fair-trade-oriented word of mouth (here recommendation intention captured by the Net Promoter Score), is introduced and discussed. Next, in the "Research Methodology" section, the measurement instruments and sample are outlined, followed by a discussion of the research findings. Finally, the paper closes with a discussion regarding research and managerial implications as opportunities to develop strategies that aim to form the basis of a structured understanding of perceived value and related consumer behavior in the context of ethical and green consumption. In sum, the main findings of this study underline that customers who reveal a high luxury perception of Fair Trade products are strengthened in the Fair Trade idea which results in higher customer perceived value as well as the willingness to recommend Fair Trade products.

2 Conceptualization and Hypotheses Development

Following a definition of sustainability as an approach to "*meet the needs of the current generation without compromising the ability of future generations to meet their own needs*" (World Commission on Environment and Development 1987) based on attitudes such as durability, quality, and timelessness, inherent synergies to the concept of luxury become apparent (Kapferer and Bastien 2009). Instead of being two opposing perspectives, a closer look reveals that sustainability and luxury share common ideals (Cvijanovich 2011). Similar to sustainable goods, luxury is managed based on a long-term perspective, the respect for rare resources and authentic craftsmanship (Janssen et al. 2013; Cvijanovich 2011; Kapferer 2010). Existing literature suggests that sustainable and green consumption is determined by a whole variety of factors. For the purposes of this study, to advance the current understanding of customer value perceptions and effective behavior in view of Fair Trade products, we examine an integrated conceptual framework, as illustrated in Fig. 1. Fair Trade products have been chosen, reasoning that those specific product categories represent both sustainability (e.g., environmental protection) and luxury aspects (e.g., exclusivity).

The key constructs and related hypotheses are described below.

Fair-trade-oriented Luxury Perception: Traditionally, luxury products can be defined as those whose price and quality ratios are the highest in the market (McKinsey 1990), and even if the ratio of functionality to price might be low, the ratio of intangible and situational utility to price is comparatively high (Nueno and Quelch 1998). The concept of luxury is used in diverse industry sectors and refers

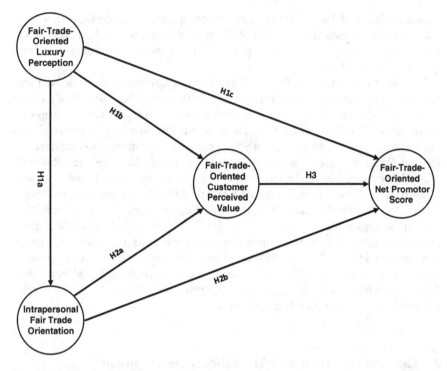

Fig. 1 Conceptual model

"*to a specific tier of offer in almost any product or service category*" (Dubois et al. 2005). In this context, and reasoning that the main factors distinguishing luxury from non-luxury products are the psychological benefits for the owner (Nia and Zaichkowsky 2000), it is reasonable to suggest that sustainable and green products can adapt a luxury strategy as well. If sustainable consumption is understood as the acceptance of limited availability and scarcity where consumers decide to buy less and better (Dubois and Paternault 1995), the connection to luxury products that are purchased based on exclusivity, durability, and rarity becomes evident. In order to analyze consumer perception and behavior with regard to sustainable and green products, we suggest that consumers who associate Fair Trade products with luxury qualities such as exclusivity and preciousness will have a positive attitude (*intrapersonal Fair Trade orientation*) and value perception of Fair Trade products (*fair-trade-oriented customer perceived value*) as well. Besides, we propose that consumers who value Fair Trade products will be more inclined to recommend the idea of Fair Trade to relevant others (*fair-trade-oriented Net Promotor Score*). Therefore, we hypothesize:

H_{1a} Fair-trade-oriented luxury perception has a positive effect on intrapersonal Fair Trade orientation.

H_{1b} Fair-trade-oriented luxury perception has a positive effect on fair-trade-oriented customer perceived value.

H_{1c} Fair-trade-oriented luxury perception has a positive effect on fair-trade-oriented Net Promotor Score.

Intrapersonal Fair Trade Orientation: Buying environmentally friendly and fairly traded products are considered to be the two most typical examples of ethical buying behavior (Shaw and Shiu 2002; Shaw and Newholm 2002; Shaw et al. 2005). As manifestation of ethics in consumerism, Fair Trade *"aims to improve the position of poor and disadvantaged food producers in the Third World by helping them to become more advantageously involved in world trade, in many developed countries"* (Jones et al. 2004, p. 77). Based on socially responsible practices and a marketing system that *"bridges artisans' needs for income, retailers' goals for transforming trade, and consumers' concerns for social responsibility through a compatible, non-exploitive, and humanizing system of international exchange"* (Littrell and Dickson 1999, p. 4). Fair Trade organizations (FTOs) pay fair wages, provide safe working conditions, are environmentally friendly, offer training, and contribute to community development (Fairtrade 2011). Consumers of Fair Trade products feel responsibility toward society and demonstrate their feelings through their purchase behavior (De Pelsmacker et al. 2005b). Existing study insights on attitudes and behaviors of Fair Trade consumers reveal that consumer motivations and their degree of intensity and loyalty toward Fair Trade products vary (e.g., Shaw and Clarke 1999; Cherrier 2007; Newholm and Shaw 2007; Low and Davenport 2007). As shown in our conceptual model, the individual Fair Trade orientation understood as consumers' psychological and behavioral responses to Fair Trade products is suggested to have a positive effect on customer value perception (*fair-trade-oriented customer perceived value*) and the willingness to recommend Fair Trade products to others (*fair-trade-oriented Net Promotor Score*):

H_{2a} Intrapersonal Fair Trade orientation has a positive effect on fair-trade-oriented customer perceived value.

H_{2b} Intrapersonal Fair Trade orientation has a positive effect on fair-trade-oriented Net Promotor Score.

Fair-trade-oriented Customer Perceived Value: As a context-dependent (MBHolbrook 1994; Parasuraman 1997), highly personal, and multidimensional concept, customer perceived value creates a trade-off between product-related benefits and sacrifices, expected by current as well as potential customers in different phases of the purchase process (Woodruff 1997; Sweeney and Soutar 2001; Graf and Maas 2008). In an attempt to examine a customer's perceived preference for and evaluation of a certain Fair Trade product, the construct of *customer perceived value* understood as *"a consumer's overall assessment of the utility of a product based on perceptions of what is received and what is given"* (Zeithaml 1988, p. 14) based on *"an interactive relativistic consumption preference experience"* (MBHolbrook 1994) is of particular importance. Concerning various sources of the customer's Fair Trade perception, a wide range of relevant present and potential motives that might explain why consumers choose to buy or avoid a certain product should be taken into consideration (De Ferran and Grunert 2007; Renard 2003; Sheth et al. 1991). According to the customer perceived value framework by Sweeney and Soutar (2001) and incorporating the findings of the meta-analysis of value perceptions' research by Smith and Colgate (2007), well-known consumption values, which are commonly divided into the four types *affective, economic, functional*, and *social*, seem to drive Fair-Trade-oriented purchase attitude and behavior (De Pelsmacker et al. 2005a). Hence, globally perceived relevance, necessity, and meaningfulness of Fair Trade products, represented in our conceptual model by the Fair-Trade-oriented customer perceived value, are expected to have a positive effect on the willingness to recommend the idea of Fair Trade to relevant others (*fair-trade-oriented Net Promotor Score*):

H₃ Fair-trade-oriented customer perceived value has a positive effect on fair-trade-oriented Net Promotor Score.

Fair-trade-oriented Net Promoter Score: Dedicated to the question of how to measure the customer's loyalty to companies, Reichheld (2003) introduced the Net Promoter Score (NPS) in Harvard Business Review as "*The One Number You Need to Grow*": He claims that the word-of-mouth metric is the single most reliable indicator for sustainable firm revenue growth as only convinced consumers would recommend a company or, respectively, its products and services (Reichheld 2003, 2006a, b). Based on survey responses to one single question, asking for the likelihood to recommend on an 11-point scale ranging from 0 to 10, the firm's Net Promoter Score was designed to measure the number of people who are likely to provide positive comments (called "*promoters*"), rating the firm a 9 or 10, minus those likely to give negative comments (called "*detractors*"), rating the firm a 6 or less. Compared to models that refer on data from multiple survey items to predict firm growth, Reichheld (2006c) reported that NPS "*yields slightly less accurate predictions for the behavior of individual customers, but a far more accurate estimate of growth for the entire business*" and a 12-point increase in NPS leads to a doubling of a company's growth rate on average according to a study by Bain & Company (Reichheld 2006a).

Nevertheless, NPS has been severely critiqued in terms of its predictive validity in relation to company growth: "*The simple truth, however, is that these claims remained largely untested by the scientific community*" (Keiningham et al. 2008, pp. 82/83). In a longitudinal study of company performance and NPS measures across industries, Keiningham et al. (2007b) were unable to replicate Reichheld's (2003) findings. Another study, conducted by Keiningham et al. (2007a), showed that a multiple-item measure gave a better prediction of retention and recommendation so that Sharp (2008) criticizes NPS even as "*fake science*" (p. 30).

Reasoning the scientific discussion on the effective added value of the NPS, the present study aims to examine the relevant drivers of NPS as the indicator for consumer's willingness to recommend a certain Fair Trade product instead of focusing on possible outcomes such as company growth. In particular, we assume that the applied fair-trade-oriented Net Promoter Score is driven by the already mentioned fair-trade-oriented luxury perception, the intrapersonal Fair Trade orientation, and the fair-trade-oriented customer perceived value.

3 Research Methodology

3.1 Questionnaire

Reflective global scales were developed and used to measure fair-trade-oriented luxury perception, intrapersonal Fair Trade orientation, and fair-trade-oriented customer perceived value. With specification to a Fair Trade context, all global items were rated on seven-point semantic differentials. Furthermore, a standard Net Promotor Score (NPS) measure as our chosen Fair Trade behavior performance success indicator was used by asking the original question *"How likely are you to recommend Fair Trade products to a friend, colleague or relative?"* on a scale from 0 (*not at all likely*) to 10 (*extremely likely*). The first version of the questionnaire was face-validated using exploratory and interviews with marketing experts from the field of business practice ($n = 5$) and science ($n = 5$), each with level of knowledge concerning sustainability aspects, as well as actual consumers of Fair Trade products ($n = 10$) to ensure an appropriate survey length and structure as well as the clarity of the items used.

3.2 Sample and Procedure

We conducted a Web-based survey in Germany to investigate the research model. In fact, Germany is a highly developed, competitive, and still growing market for sustainable products in general and Fair Trade products in particular with a high share of people demanding and consuming those products. Effectively, a strong increase in the consumption of Fair Trade products has been recorded for Germany with hitting a turnover over half a billion Euro in 2012 and indicating a growth of 33 % within one year (Fairtrade International 2013). Interviewees were recruited using the ethical and organic consumer research panel "green consumer" (this specific panel provides subjects who acknowledge to buy and consume various types of sustainable products on a regular basis) by *Toluna*, one of the leading global online market research panel companies. Survey participation was limited to consumers who had bought Fair Trade products during the last six months. In January 2013, after Fair Trade became really a mass market in Germany according to the statistics as noticed above, a total of 320 valid questionnaires were received. We used a quota sampling to select participants and to receive a well-balanced representation of any gender and age group referring to a Fair Trade-relevant target

Table 1 Demographic profile of the sample

Variable		n	%
Age	20–29 years	80	25.0
	30–39 years	80	25.0
	40–49 years	80	25.0
	50–59 years	80	25.0
Gender	Male	160	50.0
	Female	160	50.0
Marital status	Single	150	46.9
	Married	135	42.2
	Divorced	28	8.8
	Widowed	3	0.9
	No answer	4	1.3
Education	Not graduated from high school	1	0.3
	Lower secondary school	21	6.6
	Intermediate secondary school	116	36.3
	A-levels	89	27.8
	University degree	92	28.8
	No answer	1	0.3
Occupation	Full time	181	56.6
	Part-time	51	15.9
	Pensioner/retiree	12	3.8
	House wife/husband	24	7.5
	Job training	7	2.2
	Student	20	6.3
	Scholar	1	0.3
	Seeking work	23	7.2
	No answer	1	0.3
Household income	500 EUR or less	10	3.1
	501 EUR–1000 EUR	39	12.2
	1001 EUR–1500 EUR	49	15.3
	1501 EUR–2000 EUR	59	18.4
	2001 EUR–3000 EUR	89	27.8
	3001 EUR–4000 EUR	45	14.1
	4000 EUR or more	28	8.8
	No answer	1	0.3

group. Specifically, in our sample, the gender ratio was 50 % female and 50 % male with an age share of 25 % each for the sub-age group 20–29 years, 30–39 years, 40–49 years, and 50–59 years. The sample characteristics are shown in Table 1.

4 Results and Discussion

4.1 Analysis Technique

For examining the drivers and outcomes of a luxury-driven Fair Trade behavior and to check the postulated hypotheses, a partial least squares (PLS) structural modeling was employed in this exploratory study. PLS was considered as the appropriate method due to the facts that this analysis technique (a) places minimal specifications on sample sizes, (b) is suitable when the research objective involves theory building, (c) does not require a normal distribution of the manifest variables, and (d) evaluates the predictive power of the causal model. For that reason, we employed the software statistics package SmartPLS 2.0 (Ringle et al. 2005) with casewise replacement and a bootstrapping procedure (probing individual sign changes) to carry out the empirical PLS regression analysis. As illustrated in Fig. 1, the investigated PLS path model includes the reflective constructs of fair-trade-oriented luxury perception, intrapersonal Fair Trade orientation, fair-trade-oriented customer perceived value, and a fair-trade-oriented NPS measure. The result assessment is presented in the subsequent sections with reference to an outer and inner model discussion.

4.2 Evaluation of the Measurement Models

The manifest variables that are indicators for the measurement models of fair-trade-oriented luxury perception, intrapersonal Fair Trade orientation, and fair-trade-oriented customer perceived value are presented in Table 2. With regard to the assessment of the reflective measures as shown in Table 3, all factor loadings are statistically significant and well exceed the recommended value of 0.7, thus suggesting item reliability (Carmines and Zeller 1979; Hulland 1999).

Table 2 Manifest variables of the reflective measurement models

Fair-trade-oriented luxury perception	
LX_global_01	Not at all exclusive–extremely exclusive
LX_global_02	Not at all first-class–extremely first-class
LX_global_03	Not at all extravagant–extremely extravagant
LX_global_04	Not at all precious–extremely precious
Intrapersonal Fair Trade orientation	
FTO_economic_01	Not at all fair-trade-oriented–extremely fair-trade-oriented
FTO_economic_02	Not at all interested–extremely interested
Fair-trade-oriented customer perceived value	
CPV_global_01	Not at all relevant–extremely relevant
CPV_global_02	Not at all necessary–extremely necessary
CPV_global_03	Not at all meaningful–extremely meaningful

Table 3 Assessing the measurement models

	Factor loadings	Average variance explained (AVE) (%)	Cronbach's alpha	Composite reliability	Fornell–Larcker criterion (AVE > Corr^2)
Fair-trade-oriented luxury perception	0.790–0.876	69	0.854	0.900	0.69 > 0.38
Intrapersonal Fair Trade orientation	0.902–0.930	84	0.809	0.912	0.84 > 0.50
Fair-trade-oriented customer perceived value	0.888–0.908	81	0.880	0.926	0.81 > 0.58
Fair-trade-oriented Net Promotor Score	1.000	100	1.000	1.000	1.00 > 0.58

In addition, all reflective constructs exhibit satisfactory values in terms of internal consistency (Bagozzi and Yi 1988; Nunnally and Bernstein 1994): The average variance extracted (AVE) estimates range from 69 to 81 %, the Cronbach's alphas range from 0.809 to 0.880, and the composite reliability values range from 0.900 to 0.926. Moreover, the Fornell–Larcker criterion was used to assess discriminant validity (Fornell and Larcker 1981). In this study, each latent variable passes the criterion requirements, thereby satisfying discriminant validity.

4.3 Evaluation and Discussion of the Structural Model

In a next step, the predictive quality of our introduced conceptual model was assessed after ensuring the reliability and validity of each measure. Therefore, the coefficients of determination of the endogenous latent variables (R^2) were first examined. As presented in Fig. 2, the calculated R^2 values are ranging from 0.38 to 0.65. In accordance with Chin (1998), these values reveal a moderate to substantial performance. Furthermore, Stone–Geisser Q-square values were assessed (Tenenhaus et al. 2005) using a blindfolding procedure (cross-validated redundancy). All Q-square values are clearly greater than zero with 0.32 being the smallest one, as shown in Table 4. According to the empirical results, the PLS structural equation model demonstrates a high predictive relevance.

4.4 Testing the Hypotheses

A nonparametric bootstrapping procedure (320 cases and 1600 subsamples; individual sign changes) was applied to assess the significance of the path coefficients

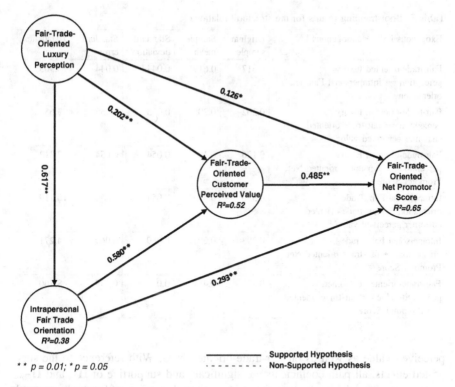

Fig. 2 Structural model

Table 4 Assessing the structural model

Endogenous LV	R^2	Q^2
Intrapersonal Fair Trade orientation	0.380	0.318
Fair-trade-oriented customer perceived value	0.521	0.411
Fair-trade-oriented Net Promotor Score	0.650	0.649

and to test the proposed hypotheses. Table 5 presents the following insights with reference to our initial hypotheses.

H_{1a} to H_{1c} *are confirmed.* The first set of research hypotheses H_{1a} to H_{1c} focused on the effect of fair-trade-oriented luxury perception on the subsequent latent variables: intrapersonal Fair Trade orientation, fair-trade-oriented customer perceived value, and fair-trade-oriented NPS. In detail, the results reveal significant positive impacts of fair-trade-oriented luxury perception on each latent variable. Thus, the empirical data provide full support for H_{1a}, H_{1b}, and H_{1c}.

H_{2a} *and* H_{2b} *are confirmed.* The second set of hypotheses postulates an influence of intrapersonal Fair Trade orientation on fair-trade-oriented customer

Table 5 Bootstrapping results for the structural relations

Exogenous LV → endogenous LV	Original sample	Sample mean	Standard deviation	Standard error	T statistics
Fair-trade-oriented luxury perception → intrapersonal Fair Trade orientation	0.617	0.619	0.044	0.044	13.903
Fair-trade-oriented luxury perception → fair-trade-oriented customer perceived value	0.202	0.205	0.055	0.055	3.652
Fair-trade-oriented luxury perception → fair-trade-oriented Net Promotor Score	0.126	0.125	0.054	0.054	2.312
Intrapersonal Fair Trade orientation → fair-trade-oriented customer perceived value	0.580	0.575	0.054	0.054	10.719
Intrapersonal Fair Trade orientation → fair-trade-oriented Net Promotor Score	0.293	0.291	0.069	0.069	4.271
Fair-trade-oriented customer perceived value → fair-trade-oriented Net Promotor Score	0.485	0.486	0.059	0.059	8.200

perceived value as well as on fair-trade-oriented NPS. With reference to the suggested effects, all path coefficients are significant and supportive of H_{2a} and H_{2b}.

H_3 is confirmed. Finally, hypothesis H_3 proposed an effect of fair-trade-oriented customer perceived value on Fair-Trade-oriented NPS. Accenting the critical role of customer perceived value as a key element for a successful product management and in support of H_3, the PLS-based results reveal a strong and significant influence of fair-trade-oriented customer perceived value on fair-trade-oriented NPS.

Taken all together, the findings stress that customers with a high luxury perception regarding Fair Trade products show a higher level of involvement and reassurance toward the idea of Fair Trade. As a consequence thereof, those customers reveal a higher level of customer perceived value of Fair Trade products (e.g., product quality) as well as the willingness to recommend (word of mouth) Fair Trade products. However, the current study only captured a single indicator, specifically the well-known Net Promoter Score, to assess the customer behavior toward Fair Trade products. In reality, customer behavior is more multifaceted. That said, the current results can only be regarded concerning the loyalty potential, but less with reference to other important behavior aspects like customer's willingness to pay a higher price (price premium) or cross-marketing potential (e.g., to buy products from other product categories).

5 Conclusion

5.1 Scientific Contribution

The first contribution of the research presented here is to provide a conceptual framework of value-based drivers of Fair Trade product perception against the backdrop of the luxury concept. Second, the empirical findings contribute to the understanding of consequences of fair-trade-based luxury perception. Customers who reveal a high luxury perception of Fair Trade products are strengthened in the Fair Trade idea which results in higher customer perceived value as well as the willingness to recommend Fair Trade products. Third, the data analysis of the applied PLS-SEM approach demonstrates that positive Fair Trade behavior is influenced by direct and indirect effects. With reference to the conducted study, being an active promoter of Fair Trade products is directly determined by the customers' perceived product value (benefit in relation to cost), but also directly and indirectly affected by the overall luxury product perception as well as the intrapersonal level of Fair Trade orientation.

5.2 Next Research Steps

For further research, our introduced conceptual model should be extended in two ways. First, the main value-based drivers fair-trade-oriented luxury perception as well as intrapersonal Fair Trade orientation should be refined in an expanded quantitative level of detail (e.g., hedonism driving luxury perception or social identification driving Fair Trade orientation), since both dimensions have a limited explanatory scope due to their global characteristics. Second, the impact of fair-trade-oriented luxury perception and intrapersonal Fair Trade orientation on a refined Fair Trade product perception (e.g., image or satisfaction) as well as Fair Trade product behavior (e.g., loyalty or price premium) should be explored. A model extension would provide managers with valuable information if and how the concept of luxury in relation to Fair Trade products enhances product's key performance indicators such as buying intention and actual behavior.

5.3 Managerial Implications

From a managerial perspective, our study may form the basis of a structured understanding of perceived value and related consumer behavior in the context of ethical and green consumption. Our results reveal that consumers who relate the concept of luxury based on aspects such as exclusivity, rarity, and authenticity with the idea of sustainability and Fair Trade orientation perceive higher value and are more inclined to buy and recommend Fair Trade products. Therefore, a comprehensive management approach dedicated to the creation and maintenance of a successful sustainable brand with lasting competitive advantage chases away any *"granola-and-hemp clichés"* (McMillan 2012) and stresses the value components of exclusivity, preciousness, and longevity. In this context, one of the key challenges of a sustainable communication approach is to raise and sustain the public awareness of existing environmental and social concerns and keep generations of consumers informed about the exploitation of nature and producers as well as unfair trading practices. However, instead of being a self-proclaimed moralizer and accuser, marketing managers have to provide viable alternatives that can help to overcome these problems. In the trade-off between conventional and ethical products, present and future consumers are only willing to accept these products as sustainable opportunities when the perception of deeper values is satisfactory. As shown in this study, the sustainability sector can learn from the management of luxury brands where a rich history of value-based marketing exists and has been successful for generations of consumers.

References

Bagozzi RP, Yi Y (1988) On the evaluation of structural equation models. J Acad Mark Sci 16 (1):74–94

Browne A, Harris P, Hofney-Collins A, Pasiecznik N, Wallace R (2000) Organic production and ethical trade: definition, practice and links. Food Policy 25(1):69–89

Carmines EG, Zeller RA (1979) Reliability and validity assessment. Sage Publications, Newbury Park

Cherrier H (2007) Ethical consumption practices: co-production of self-expression and social recognition. J Consum Behav 6(5):321–335

Chin WW (1998) The partial least squares approach to structural equation modelling. In: Marcoulides GA (ed) Modern methods for business research. Lawrence Erlbaum Associates, Mahwah, pp 295–358

Cowe R, Williams S (2000) Who are the ethical consumers?. The Cooperative Bank, London

Creyer E, Ross W (1997) The influence of firm behavior on purchase intention: do consumers really care about business ethics? J Consum Mark 14(6):421–433

Cvijanovich M (2011) Sustainable luxury: Oxymoron? In: Lecture in luxury and sustainability. http://www.mcmdesignstudio.ch/files/Guest%20professor%20Lucern%20School%20of%20Art%20%20and%20Design.pdf. Accessed 15 Jan 2015

De Ferran F, Grunert KG (2007) French fair trade coffee buyers' purchasing motives: an exploratory study using means-end chains analysis. Food Qual Prefer 18(2):218–229

De Pelsmacker P, Driesen L, Rayp G (2005a) Do consumers care about ethics? Willingness to pay for fair-trade coffee. J Consum Aff 39(2):363–385

De Pelsmacker P, Janssens W, Sterckx E, Mielants C (2005b) Consumer preferences for the marketing of ethically labelled coffee. Int Mark Rev 22(5):512–530

Dubois B, Czellar S, Laurent G (2005) Consumer segments based on attitudes toward luxury: empirical evidence from twenty countries. Mark Lett 16(2):115–128

Dubois B, Paternault C (1995) Observations: understanding the world of international luxury brands: the dream formula. J Advertising Res 35(4):69–75

Elkington J, Hailes J (1989) The green consumer's supermarket shopping guide. Gollancz, London

Fairtrade (2011) What is fairtrade. Available via Fairtrade Foundation. http://www.fairtrade.org. uk/about_what_is_fairtrade.htm. Accessed 13 Jan 2015

Fairtrade International (2013) Unlocking the power. Annual Report 2012–13. Available via Fairtrade Foundation. http://www.fairtrade.net/fileadmin/user_upload/content/2009/resources/ 2012-13_AnnualReport_FairtradeIntl_web.pdf. Accessed 27 Apr 2015

FLO (2010) Fairtrade at a glance. Available via FLO. http://www.fairtrade.net/fileadmin/user_ upload/content/2009/resources/2011-05_fairtrade_at_a_glance-en.pdf. Accessed 13 Jan 2015

Fornell C, Larcker DF (1981) Evaluating structural equation models with unobservable variables and measurement error. J Mark Res 18(2):39–50

Graf A, Maas P (2008) Customer value from a customer perspective: a comprehensive review. Journal für Betriebswirtschaft 58(1):1–20

Holbrook MB (1994) The nature of customer value: an axiology of services in the consumption experience. In Rust R, Oliver RL (eds) Service quality. New directions in theory and practice. Thousand Oaks, Sage Publications, pp 21–71

Hulland J (1999) Use of partial least squares (PLS) in strategic management research: a review of four recent studies. Strateg Manag J 20(2):195–204

Janssen C, Vanhamme J, Lindgreen A, Lefebvre C (2013) The catch-22 of responsible luxury: effects of luxury product characteristics on consumers' perception of fit with corporate social responsibility. J Bus Ethics 119(1):45–57

Jones P, Comfort D, Hillier D (2004) Developing customer relationships through fair trade: a case study from the retail market in the UK. Manag Res News 27(3):77–87

Kapferer J-N (2010) All that glitters is not green: the challenge of sustainable luxury. EBR (November/December):40–45

Kapferer J-N, Bastien V (2009) The specificity of luxury management: turning marketing upside down. J Brand Manag 16(5/6):311–322

Kates RW, Parris TM, Leiserowitz AA (2005) What is sustainable development: goals, indicators, values, and practice. Environment 47(3):8–21

Keiningham TL, Cooil B, Aksoy L, Andreassen TW, Weiner J (2007a) The value of different customer satisfaction and loyalty metrics in predicting customer retention, recommendation, and share-of-wallet. Manag Serv Q 17(4):361–384

Keiningham TL, Cooil B, Andreassen TW, Aksoy L (2007b) A longitudinal examination of net promoter and firm revenue growth. J Mark 71(3):39–51

Keiningham TL, Aksoy L, Cooil B, Andreassen TW, Williams L (2008) A holistic examination of net promoter. J Database Mark Customer Strategy Manag 15(2):79–90

King MF, Bruner GC (2000) Social desirability bias: a neglected aspect of validity testing. Psychol Mark 17(2):79–103

Littrell MA, Dickson MA (1999) Social responsibility in the global market: fair trade of cultural products. Sage Publications, Thousand Oaks

Low W, Davenport E (2007) To boldly go... exploring ethical spaces to re-politicise ethical consumption and fair trade. J Consum Behav 6(5):336–348

McKinsey (1990) The luxury industry: an asset for France. McKinsey, Paris

McMillan A (2012) 7 sustainable luxury brands making eco-friendly fashion. Available via Elle Canada, April 2012. http://www.ellecanada.com/fashion/trends/7-sustainable-luxury-brands- making-eco-friendly-fashion/a/57121. Accessed 13 Jan 2015

Mohr LA, Webb DJ, Harris KE (2001) Do consumers expect companies to be socially responsible? The impact of corporate social responsibility on buying behavior. J Consum Aff 35(1):45–72

Newholm T, Shaw D (2007) Studying the ethical consumer: a review of research. J Consum Behav 6(5):253–270

Nia A, Zaichkowsky J (2000) Do counterfeits devalue the ownership of luxury brands? JPBM 9 (7):485–497

Nicholls AJ (2002) Strategic options in fair trade retailing. Int J Retail Distrib Manag 30(1):6–17

Nueno JL, Quelch JA (1998) The mass marketing of luxury. Bus Horiz 41(6):61–68

Nunnally J, Bernstein I (1994) Psychometric theory, 3rd edn. McGraw Hill, New York

Papaoikonomou E, Ryan G, Ginieis M (2011) Towards a holistic approach of the attitude behaviour gap in ethical consumer behaviours: empirical evidence from Spain. Int Adv Econ Res 17(1):77–88

Parasuraman A (1997) Reflections on gaining competitive advantage through customer value. J Acad Mark Sci 25(2):154–161

Peet R, Watts M (2002) Liberation ecology: development, sustainability, and environment in an age of market triumphalism. In: Peet R, Watts M (ed) Liberation ecologies: environment, development, social movements. Routledge, New York, pp 1–45

Phau I, Prendergast G (2000) Consuming luxury brands: The relevance of the 'rarity principle'. J Brand Manag 8(2):122–138

Reichheld FF (2003) The one number you need to grow. Harvard Bus Rev 81(12):46–54

Reichheld FF (2006a) The microeconomics of customer relationships. MIT Sloan Manag Rev 47 (2):73–78

Reichheld FF (2006b) The ultimate question: driving good profits and true growth. Harvard Business School Press, Boston

Reichheld FF (2006c) Questions about NPS-and some answers. http://netpromoter.typepad.com/fred_reichheld/2006/07/questions_about.html. Accessed 18 Jan 2015

Renard M-C (2003) Fair trade: quality, market and conventions. J Rural Stud 19(1):87–96

Ringle CM, Wende S, Will A (2005) SmartPLS 2.0 M3. http://www.smartpls.de. Accessed 18 Feb 2014

Rosenbaum M (1993) Trading standards: will customers now shop for fair play?: Group plans to flag products that are third world-friendly. The independent, June 6, 1993. http://www.independent.co.uk/news/business/trading-standards-will-customer-now-shop-for-fair-play-group-plans-to-flag-products-that-are-third-worldfriendly-1489932.html. Accessed 18 Jan 2015

Sharp B (2008) Net promoter score fails the test. Market research buyers beware. Marketing Research, Winter 2008. http://www.keepandshare.com/doc/1857900/net-promoter-score-fails-the-test-pdf-april-14-2010-4-38-pm-285k?dn=y&dnad=y. Accessed 18 Feb 2014

Shaw D, Clarke I (1999) Belief formation in ethical consumer groups: an exploratory study. J Mark Intell Plann 17(2):109–120

Shaw D, Newholm T (2002) Voluntary simplicity and the ethics of consumption. J Psychol Mark 19(2):167–185

Shaw D, Shiu E (2002) The role of consumer obligation and self-identity in ethical consumer choice. Int J Consum Stud 26(2):109–116

Shaw D, Grehan E, Shiu E, Hassan L, Thomson J (2005) An exploration of values in ethical consumer decision making. J Consum Behav 4(3):185–200

Sheth JN, Newman BL, Gross BL (1991) Why we buy what we buy: a theory of consumption values. J Bus Res 22(2):159–170

Smith JB, Colgate M (2007) Customer value creation: a practical framework. J Mark Theor Pract 15(1):7–23

Strong C (1996) Features contributing to the growth of ethical consumerism-a preliminary investigation. Mark Intell Plann 14(5):5–13

Sweeney JC, Soutar GN (2001) Consumer-perceived value: the development of a multiple item scale. J Retail 77(2):203–220

Tenenhaus M, Vinzi VE, Chatelin Y-M, Lauro C (2005) PLS path modelling. Comput Stat Data Anal 48(1):159–205

Woodruff R (1997) Customer value: the next source for competitive advantage. J Acad Mark Sci 25(2):139–153

World Commission on Environment and Development (1987) Our common future. Bruntland Report, Oxford University Press, Oxford

Zeithaml VA (1988) Consumer perceptions of price, quality and value: a means-end model and synthesis of evidence. J Mark 52(3):2–22

The Sustainable Luxury Craft of Bespoke Tailoring and Its' Enduring Competitive Advantage

WenYing Claire Shih and Konstantinos Agrafiotis

Abstract Luxury's attraction from the hedonistic customs of ancient times to the latest craze for the season's hot luxury bag has always characterized humanity's consumption behavior. The contemporaneous condition of global luxury markets has led luxury brands into an unprecedented proliferation of products, as customers especially in emerging economies crave for the latest luxury items. The democratization phenomenon of mainstream fashion has also enveloped the luxury sector in a wasteful consumption pattern of global proportions which does not bode well in sustainability terms. Indeed, the rapid depletion of the Earth's natural resources indicates that both the fashion and the luxury sector are culprits in this environmental deterioration. Remedial actions are urgently needed, and the deceleration of fashion and luxury consumption is one of the suggested strategies. Deceleration combined with competitiveness may seem to be a perplexing task for luxury companies to tackle; however, there is an example in the luxury sector where slowness and competitiveness not only coexist but also thrive. The bespoke tailors in Savile Row, London, have demonstrated that competitive advantage can be achieved in a dense network of cooperation and competition. The authors have employed the relational view of the firm to identify that network resources can be applied to the slow tailoring craft and constitute a competitive strategy.

Keywords Sustainable luxury · Bespoke tailoring · Slow fashion movement · Network resources · Competitiveness

W.C. Shih (✉)
Department of Fashion Design, Hsuan Chuang University, Hsinchu, Taiwan
e-mail: wenyingshih@yahoo.com

K. Agrafiotis
Independent Fashion Business Consultant, London, UK

© Springer Science+Business Media Singapore 2016
M.A. Gardetti and S.S. Muthu (eds.), *Handbook of Sustainable Luxury Textiles and Fashion*, Environmental Footprints and Eco-design of Products and Processes, DOI 10.1007/978-981-287-742-0_8

1   Introduction

The lure of luxury consumption has always been a contentious characteristic of the human race. The hedonistic customs of the aristocracy of the Athenian Democracy and the Epicurean philosophy principles of pursuing uncomplicated luxuries in life were possibly among the first luxury activities recorded and criticized in the ancient civilized world (De Botton 2000; Berry 1994).

In the late nineteenth century, Veblen (1994) observed the ostentatious behavior of the *nouveaux riches* in his "conspicuous consumption" model. During the following decades, nascent luxury crafts appeared mainly in the European continent to cater for the desires of the *haute bougeoisie* of industrialists who demanded status symbols. These luxury crafts evolved into independent luxury houses and consequently brands by serving the world's elites (Hilton 2009).

Luxury conglomeration, a 1990s phenomenon, referred to the end of independence for the majority of luxury houses as they were to become acquisition targets of powerful luxury groups which also emerged during the same period. Transformation into global brands followed, for luxury groups expansion plans correlated well with the globalization of markets and of taste (Agins 2000).

The democratization of fashion in the 1990s facilitated by new technologies and a massive retail expansion of global proportions ushered mid-level brands into an unprecedented proliferation of ranges produced at the speed of light and offered to the unsuspecting *fashionistas* (Cline 2013). Luxury brands also responded in the same fashion cashing in customers insatiable appetite for luxury objects especially in emerging economies (Cline 2013).

This acceleration of fashion consumption may lead to an untenable future, as the depletion of natural resources has already violated some of the Earth's sustainability thresholds. The same holds true for luxury, which has not looked into sustainable practices until recently. Moreover, as the World Wide Fund (WWF) report has revealed, luxury brands score very low in terms of sustainability (Bendell and Kleanthous 2007). In broad fashion terms, the damage inflicted to the environment can be found in two main facts. The clothing production consumes vast amounts of energy, from freshwater to irrigate cotton plantations, and water used in textile dyeing mills and industrial laundering facilities as well as domestic washing. Fossil fuels generate energy for the above uses and thus result in air and water pollution (Cline 2013). The other fact is the wasteful consumption where the ever-shortening life cycle of fashion products leads to their disposal as rubbish to landfills even before the end of a season (Siegle 2011; Fletcher 2008). Disposal to landfills as far as luxury items are concerned may be not the preferred option of luxury's customers. However, this sector needs to address its responsibilities for responsible and ethical production as well as to redefine the objective of luxury consumption by expressing deeper values. In other words, luxury is of reorientation strategies in a world of finite resources from diamonds to exotic leathers and provocative displays of wealth, especially in emerging economies, where in most cases social inequality prevails (Bendell and Kleanthous 2007).

The deceleration of fashion consumption has been put forward as one of the principal ideas behind the slow fashion philosophy. This movement also promulgates other concepts such as quality over quantity, timelessness over meaningless trends, and recommends customers to buy less but buy better. All these concepts relate not only to mainstream fashion but also to luxury consumption (Fletcher 2008). This study aims to understand how the luxury sector sustains its current consumption momentum and simultaneously espouses the logic of sustainable development. Realistically, there are no definitive answers yet to these questions. Nevertheless, there is an example within the luxury fashion industry which has managed to sustain its competitive advantage for nearly 200 years, by demonstrating a remarkable resilience in adapting to the *zeitgeist* of every decade.

The Savile Row tailors, a British institution, have achieved the feat of redeploying a blend of unique resources in a dense network of relations. They not only survive in the luxury sector but also thrive for centuries. They have been accomplished neither by fast fashion, nor due to the proliferation of nonsensical mass luxury ranges, but by a craft that is based on exquisite sculpting skills, undisputed beauty and many hours of assembling a garment by hand. These tailors have also attracted very wealthy customers from all over the world, who are patient enough to wait for months for their suits to reach perfection. Another important parameter is that the Savile Row tailors enjoy a revival in recent time as a global resurgence of crafts is currently observed in luxury, ranging from mechanical watches to clothing, bags, and shoes.

The theoretical framework of the relational view of the firm, an extension of the resource-based view, is employed in the chapter, for it explains the competitive advantage of a company operating within a network. Network resources and network sourcing are crucial to the Savile Row tailors since they make extensive use of them in their practices. Geographical proximity is also imperative to both network resources and sourcing, because cloth and haberdashery merchants together with the dexterous alteration tailors are all located in the vicinity of tailor houses. Consequently, this dense network of resources and relations has been reinforcing its participant firms for centuries and will possibly continue to do so in the future within the sustainable development mindset.

The remainder of the chapter is organized in the following order. In section two, the authors discuss a concise chronology of cultural and societal events of humanity's affinity with the notion of luxury from the antiquity to the modern age. The current perception of luxury is also discussed together with the diversification strategies employed by the dominant luxury groups. The sustainability issue in fashion and the luxury sector is then addressed in section three, coupled with the industry's reawakening to sustainable practices and the resulting corrective actions. In Sect. 4, the authors probe into the slow fashion philosophy and combine this to the recent revival of artisanal production methods. Next, Sect. 5 explains the theory of the relational view and highlights the importance of relational networks, network resources, and dynamic capabilities as these are instrumental to competitive advantage. In Sect. 6, the research methodology and the methods are employed in order to answer the research question. In Sect. 7, the authors focus on the artisanal practices of the Savile Row tailors, discussing their relational production network and retail facades. The final section is the conclusion of this chapter.

2 Luxury: A Short Chronology

The word luxury derives from the Latin word "lux" which means light, as it radiates in opulence. The Romans extended the word to "luxus" which literally means extravagant lifestyle, sumptuousness, and the display of wealth (Dubois et al. 2005). In the traditional sense which broadly holds today, luxury has been related for centuries to the possession of status, superlative quality, and scarcity (Atwal and Williams 2009; Nueno and Quelch 1998). Luxury can be the object of desire by stimulating pleasure with its beauty. This makes a luxury item physically attractive and not only to be purchased and displayed in a conspicuous manner (Kapferer 2004; Kemp 1998).

It was the Greeks who gave luxury its meaning in their philosophical thoughts by observing and criticizing mainly the Athenian aristocracy and its luxuriating living manners (De Botton 2000). Three philosophy schools—those of Aristotle's Lyceum, the Garden of Epicurus, and the Cyrenaics represented by Aristippus—contemplating ethics, morality, happiness, and pleasure drew very different conclusions in their philosophical discourses on hedonism, which had a lasting effect on Western civilization. Aristotle (1998) criticized the extravagant individual and his/her flaunting of wealth as a tasteless *show off* believing that he/she would be admired by his/her peers. Instead, Aristotle suggested that individuals should follow the difficult paths of aspiring to values such as valor and nobility. Epicurus, on the other hand, held a different view as he believed that every individual was entitled to moderate uncomplicated pleasures. He also advocated a lack of physical pain and lack of mental afflictions since their avoidance could lead to the ultimate goals of long-lasting pleasure and happiness (Woolf 2010). The Cyrenaics espoused an alternative perspective from both Epicurean and Aristotelean thoughts. For them, hedonism and instant gratification of the body and the mind were paramount to happiness as pleasure is the ultimate *chief good* in its Socratic context (Tsouna 2007; Zilioli 2014).

The Aristotelean thinking eventually prevailed, since Christian values that had spread over the European continent imposed restrictions on luxury through sumptuary laws that decreed specific ways of social behavior by refraining from public displays of wealth. Moreover, lust, one of the seven deadly sins in Christian beliefs, was associated with *luxuria* in a negative connotation of pleasure and desire which became the dominant mindset for centuries (Berry 1994).

It was during the period of the Enlightenment that perceptions of luxury started to change as pleasure and luxury were increasingly detached from Christian moral values to more secular social norms (Berry 1994; Berthon et al. 2009). Adam Smith in his seminal treatise provided a classification of goods which still broadly holds today. The luxury goods are scarce, difficult to find and to make, and are also very expensive (Berthon et al. 2009).

As the world moved into the Industrial Age with the advent of modern cities and societies, luxury and fashion for the elites served as means of distinction from inferior social classes which in turn tried to emulate the manners of the elites by imitation (Hilton 2009). Fashion's trickle-down phenomenon observed by Simmel (1957) in his classic theory explains elitist social behavior in fashion consumption. Veblen (1994) then introduced the concept, *conspicuous consumption*, and constructed a model of social hierarchy where the rich consume luxury goods in order to confirm their status, enhance their power, and secure their inclusion into the elite class, while the *nouveaux riches* and the middle classes consume luxury in order to gain status and display wealth in a purely conspicuous manner (Truong et al. 2008). Hilton (2009) also notes that despite the advent of mass production and mass markets following the Industrial Revolution, the elites have been appalled by the cheap luxuries for the masses and sought to adhere to elitist ideals restraining luxury's democratization.

However, in an ironic twist of history, luxury in the latter part of the twentieth century followed the reverse path as luxury's prerogative to the privileged few was diffused to the masses in what is called the estheticization of commerce (Postrel 2003; Weber 2007). This phenomenon has been explained by Berthon et al. (2009) who argue that luxury perceptions have shifted from great craftsmanship and durability to marketing, and more recently to the realm of experiential luxury. Twitchell (2002) notes that the means for business growth lie in the trickle-down effect since it generates "luxury" products that the middle classes aspire to. In economic terminology, luxury products are described and consequently classified as "Veblen goods" because they do not conform to the principal of economic rationality. This principal stipulates that as the price of an item of goods rises, consumers' reaction is to buy less of it. As price increases, it stimulates more demand in terms of luxury, which may sound paradoxical, but is the prevailing economic model (Frank 2014; Kapferer and Bastien 2012).

The contemporary notion of luxury will then be discussed, since its conception and meaning have been shifted dramatically during the first decade of the twenty-first century.

2.1 The Contemporary Meaning of Luxury

Rad (2012) states that the global luxury sector in 1985 was valued at 20 billion US dollars, and this figure increased to 68 billion in 2000. Bellaiche et al. (2010) incorporated as luxury travel, hotels, and cars into the luxury industry sectors estimating 1 trillion US dollars. In a seemingly more conservative recent report conducted by Transparency Market Research (2014), the luxury sector is expected to grow at an annual rate of 3.45 % from 2014 to 2020 to reach a total value of approximately 375 billion US dollars. Whatever the above estimates present, they point to impressive growth for a sector. What transformed the industry from independent family-owned companies to global luxury brands is discussed below.

2.2   Diversification Triggers Unprecedented Growth

In the 1990s, luxury conglomeration became the name of the game in the sector. Aggressive consolidation followed where the majority of independent usually family-owned luxury houses were absorbed by a small number of mainly French luxury conglomerates. Moreover, these luxury conglomerates in cases fought bitter legal battles for acquisition rights between them (Agins 2000; Tungate 2012).

Once consolidation was put in place, the new luxury masters and their marketing wizards made use of diversification strategies extending and stretching luxury brands to an unprecedented scale and even into unrelated territories. In the current state of the luxury sector, there are two types of brand extensions practiced. The first involves the well-known pyramid-like model of vertical extensions, e.g., Chanel perfumes and sunglasses, which are priced relatively lower in comparison with other Chanel products such as clothing and jewelry. The second that of horizontal extensions involves product/service launches in often unrelated product/service segments which are equally expensive such as furnishings (Ralph Lauren wall paints and bed linen) and hotels (Armani) (Agins 2000; Kapferer and Bastien 2012; Stankeviciute and Hoffmann 2010; Tungate 2012).

More analytically, this relentless pace of diversification has led to the trickling-down effect to the masses which is now re-christened "the democratization of luxury" for it caters to consumers aspirations for designed products by famous fashion designers at reasonable prices. Examples include H&M's collaborations every season with designer or luxurious brands where the eponymous designer procures simplified versions of his/her design work in the form of a capsule collection (Atwal and Williams 2009; Cline 2013; Postrel 2003). Simultaneously, the reverse motion has also been observed within the concept described as *trading up* by Silverstein and Fiske (2003) where affluent consumers are more than willing to pay a premium price for "luxury" products which are mass produced by luxury brands (Danziger 2005; Thomas 2007; Tungate 2012; Weber 2007). Moreover, the rise of affluent classes around the world has fueled the demand for luxury, thus blurring the boundaries of what in reality qualifies as a luxury item (Husic and Cicic 2009; Weber 2007). This phenomenon is defined as *masstige* deriving from the words of mass and prestige where a luxury brand stretches downward by devising derivative sub-brands from its core concept (Agins 2000). Examples of this extension tactic abound in fashion retail markets, from Armani Jeans to CK by Calvin Klein. These extensions form the lower tiers of a designer brand and as a result they reach an aspiring middle market public (Danziger 2005; Silverstein and Fiske 2003; Stankeviciute and Hoffmann 2010). Nevertheless, it must be clarified that the "real luxury" still remains inaccessible, since the principal of rarity prevails in all product categories. It is just the fact that through the diversification strategies, luxury's trinkets became available to wider publics across the world due to economic growth especially in emerging economies (Doran 2013; Kapferer and Bastien 2012; Phau and Pendergast 2000).

The diversification of luxury has also demanded a better classification in marketing terms, since not all luxury brands refer to a single homogenous concept because brands and sub-brands occupy various positions within the luxury's stratification together with their related product categories (Agins 2000; Stankeviciute and Hoffmann 2010; Tungate 2012). In this concept, some luxury brands and consequent product categories are considered more luxurious than others as this is reflected in differing price architectures and also customer perceptions (Berthon et al. 2009; Vigneron and Johnson 2004). For example, the Hermes Kelly bag commands a higher price premium in comparison with a similar size Louis Vuitton bag. Despite the democratization which supposedly could dilute the brand value, luxury brands have managed to maintain the mystique by employing selective distribution strategies which have reinforced the characteristic of exclusivity to customers' perceptions (Doran 2013; Kapferer and Bastien 2012). Therefore, luxury brands in their brand extension strategies have achieved to engineer impressive portfolios of successful brands by keeping untouched the critical core of the parent brand (Danziger 2005; Phau and Pendergast 2000; Stankeviciute and Hoffmann 2010; Tungate 2012).

2.3 Subjectivity, Emotions, and Experiences in Luxury

Subjectivity abounds in luxury as its boundaries are inclined to individual and social perceptions and interpretations (Doran 2013; Guercini and Ranfagni 2013; Kapferer and Bastien 2012). Simply put, what may constitute a luxury for a social class and indeed a whole country can be a commodity for another country and social class (Berthon et al. 2009; Frank 2014; Vickers and Renand 2003). Moreover, luxury cannot be easily explained and thus defined, solely by its characteristics for it forms a fusion of the material in its objective context, the social in its collective context, and the individual within its subjective context (Berthon et al. 2009). In sustainability terms, Guercini and Ranfagni (2013) argue that luxury is the polar opposite of sustainability since it is superfluous, conspicuous, and excessive devoid of any utilitarian use. These facts make the term notoriously difficult to define (Doran 2013).

Other important dimension of luxury is its evocative power by stirring strong emotions during the purchasing process as human senses are stimulated. This experiential element may lead customers to dream of fantasy worlds in the sense of escapism from life's mundane activities. It is equally important to mention that these fantasy worlds are induced by retail marketing experts as customers are conditioned to participate willingly in the experiences offered and pay a premium for them as well. Thus, in its more ethereal quality, luxury branding can be viewed as an experiential process because customers are immersed into sensational product design, superb packaging, outstanding service provision, and impressive store atmospherics (Berthon et al. 2009; Pine and Gilmore 1999; Wiedmann et al. 2007).

In the next section, the sustainability issue in luxury fashion is introduced together with the corrective actions and initiatives undertaken by some foresighted luxury companies.

3 Sustainable Luxury in Fashion, Green Emerges as the New Black!

Sustainability in the fashion industry and more particularly in luxury forms a fraction of the far bigger picture of sustainable development, as this was promulgated by the United Nations (UN) landmark report, *Our Common Future* (Brundtland 1987). Twenty-five years later, the UN produced an updated version with the progress that had been achieved. Unfortunately instead of progress, the current state of the environment has been exacerbated to the extent that scientists refer to violations of a safe operating space. The reasons are numerous but the major one is human enterprise which has rapidly depleted natural resources on which humanity will depend on for survival in the medium term (United Nations 2012). Currently, the Earth's *sustainable carrying capacity* is at risk because of population growth, unethical business practices, rising affluence, and patterns of wasteful consumption, all of which have prevailed across both developed and developing countries (Smitha 2011; United Nations 2012).

3.1 Global Reawakening

Remedial actions are urgently needed and it seems that there is a movement toward the right direction in a loose formation of government agencies, scholars, environmental activist associations, journalists, and responsible businesses, who have recently raised public awareness on environmental issues and wasteful consumption patterns. The wake-up call is led by the UN, as always has been, by instigating a mindset shift through its reports on sustainability with useful recommendations and fair practices for all businesses to follow. These range from the traceability of the global supply chains (United Nations Global Compact Office 2014) to raising awareness in corporate sustainability by integrated governance, by inter-firm collaborations, and by addressing systemic issues (Kiron et al. 2015; United Nations Environment Programme 2014).

3.2   The Fast Fashion Mindset and Its Catastrophic Impact

The global clothing industry and its fast fashion business model, where the speed of production and consumption has dramatically shortened, are certainly responsible for this environmental deterioration. This fast business model is based on shortening each season's time slots between production and retail inducing customers to buy more as stores' shelves are replenished approximately every 30 days not with repeat orders but with fresh merchandise. Customers' attachment to this business model is to crave for new ranges every month because the traditional pattern of two fashion seasons is now replaced by approximately 5–6 fashion cycles of new merchandise per season. This is combined with the race to the bottom for the cheapest possible production location where fashion's *flavor of the month* has an irresistible price tag attached on it. As a result, customers are conditioned to buy more and more often and as wardrobes are filled with unwanted clothing purchased at low price points, it is natural to discard the old to make room for the new. Moreover, in the event that one may add the recent deadly accidents in garment factories then the combination becomes certainly *lethal* in sustainability terms for both the environment and human cost (Cachon and Swinney 2011; Cline 2013; Gam and Banning 2011; Rose 2014; Siegle 2011). Fast fashion examples abound in the industry from the "inventors" of the model, the retailer Zara, to H&M, Gap and Primark just to name a few. What all these retail brands have in common is sales and profits greed amplified by the blessings of economists and business analysts. This aggressive business model compels retailers' contract manufacturers to cut costs in every possible way. This is translated to the lowest common denominator in labor costs, quality, and environmental considerations since speed is the imperative concern to the detriment of all other variables in the retail equation (Cline 2013; Siegle 2011; Smitha 2011).

3.3   The Luxury Sector: A Silent Culprit in Sustainable Practices

Luxury, on the other hand, seems to be impervious to the pressures of the lowest labor costs in garment factories of least developed countries, cost economization at all times, and endless acceleration to replenish shelves and racks in retail chains. The luxury industry may have not preoccupied itself too much with the race to the bottom for the cheapest price. However, luxury's massive diversification of the past years in every direction points to the fact that the sector has been far from innocent in sustainability terms among other obscure practices (Bendell and Kleanthous 2007; Thomas 2007; Weber 2007). As an example, in March 2007, LVMH the largest luxury conglomerate of the sector was expelled from the FTSE4 Good index, which monitors businesses conforming to environmental and social criteria in supply chain-related issues. Fearing of reputational damages in the financial markets, LVMH instigated some remedial actions and re-entered the index in March 2009 (Siegle 2011).

3.4 Fashion Finally Institutes Some Corrective Actions

Amid the dismal perspectives on sustainable development, both mainstream fashion and luxury have recently taken some actions to rectify the situation. Public awareness has played its part as an increasing number of customers want to know where their clothes were made and under which conditions both environmental and labor (Clifford 2013; Friedman 2010). French customers interviewed for a luxury study have demonstrated their concerns on sustainability issues, and also they have expressed the desire that French luxury goods should be manufactured in France as this sustains craftsmanship and *rare know-how* could disappear (Kapferer 2010, 2012).

A number of clothing companies together with trade bodies and universities have also responded to sustainability demands by forming associations such as the Sustainable Apparel Coalition (SAC) in the USA (Clifford 2013). The association has developed a tool the Higg Index which is a self-assessment system of the clothing and footwear sectors that provides a comprehensive examination of a fashion product as it passes from various stages of the supply chain. This examination pertains to environmental, social, and labor impacts. In this context, the index asks its participant firms with qualitative and some quantitative questions. These are practice related and based on the answers given by a particular company the index monitors a firm's sustainability performance. The index also conditions actions for improvement for its member companies. In 2013, a new version of the index was introduced where footwear and also the important social and labor issues were included. The index currently comprises three sections: the Facility Tools which pertain to manufacturing plants, materials, and packaging as well as social and labor concerns at the production level of operations; the Brand Tools which as the name suggests are brand-specific and monitor environmental, social, and labor practices; and the Product Tools which are subdivided into the Rapid Design Module and the Materials Sustainability Index. The former guides designer decisions on prototype development. The latter is an online platform of materials scoreboards where product development teams can seek advice and also input information on new materials sustainability (Clifford 2013; Savile Row Bespoke Association 2015).

Another persisting problem is how to define a sustainable fashion product because even the UN has not yet devised a precise definition. This could possibly resolve a lot of confusion among customers and retailers alike, but there has been a feeling of reluctance in both luxury and mainstream brands. The reason for this is that fashion has always been addressing more desires than necessities and in this sense it has looked somehow hypocritical for luxury fashion to engage in environmental issues (Friedman 2010). Hypocrisies aside, Kering the luxury conglomerate has taken the initiative to compose the first "environmental profit and loss" accounting methodology in the luxury sector. In its broadest context, this methodology measures the environmental impact of Kering's operations across the supply chain, which assists the company to reduce its footprint (Daveu 2014; Jackson 2011). In the words of Kering's president: ".... It is up to us to show initiative, to be extremely proactive and go beyond simple compliance rules." Kering among other initiatives undertaken in 2014 signed a five-year partnership with the London College of Fashion lending support to fashion education on sustainability issues. It has established a laboratory in order to devise sustainable solutions in materials and manufacturing. Kering also has created a program to manage and monitor the supply and trade of exotic leathers which are widely used in luxury accessories (BOF Team 2014).

Ermenegildo Zegna, the luxury menswear brand with a textiles and tailoring history that spans over a century has instigated a sustainability program by collaborating with the luxury textile mill Loro Piana. In this collaboration, both companies have committed resources by supporting Peruvian shepherds in raising Vicuna an endangered species animal that produces wool of exceptional quality. Thus, they have sustained both the diversity of the species together with shepherds' livelihoods and also they have secured the supply of wool for their suiting cloths (Bendell 2012; Kahn 2009).

Another option currently planned by luxury businesses is the cleaning up of supply chains, but this is not an easy task since top management has realized that in order to overcome hurdles, collaborative actions come to the fore. These actions need to be adopted both internally and externally at inter-firm level (Daveu 2014; Jackson 2011).

Realistically, these initiatives regarding sustainability in luxury are fraught with challenges, are time consuming, and are not entirely integrated, and the sector is just awakening to sustainability demands. Nevertheless, the unifying force is understood by all parties to be deference for the environment. Sustainability in luxury can be viewed from a variety of different perspectives, and as a result it may generate different approaches to strategy formulation and deployment and always in accordance with the differing desires of global publics (Gardetti and Torres 2013; Kapferer 2010, 2012). As an example, the *casual luxury* concept in its more immaterial nature is more in demand in advanced mature markets while *bling* and ostentatious display is more in demand in emerging economies, such as China (Kleanthous 2011; Rambourg 2014).

In the next section, the authors focus on the recent revival of artisanal production and its connection to sustainable luxury. This is combined with the slow fashion movement which advances the idea of timelessness and quality which forms the polar opposite of the meaningless fast fashion business model.

4   Craft Revival and the Slow Fashion Philosophy

There is a broad consensus among designers of luxury brands of a revival pertaining to the classic luxury values of timelessness, durability, classic style that never dates, and superlative craftsmanship. These values have been at the core of luxury design and construction for decades. In essence, luxury business practices should have always been associated with the above-mentioned characteristics, but these have been omitted amid the diversification craze (Friedman 2010; Gardetti and Torres 2013; Kahn 2009). Nevertheless, there is a craft revival in recent time as apprentices flourish in various sectors of luxury ranging from watchmaking, to handmade lace and bespoke suits. Moreover, some luxury brands have showcased their craft practices to the public thus raising awareness of this highly skilled artisan method of crafting a unique product (Halzack 2014; Joy et al. 2012). In this concept, artisanal practices seem to differentiate authentic luxury from mass luxury as customers acknowledge the superiority of a handmade object and value it accordingly (Halzack 2014).

Artisanal manufacturing by definition is a slow process that is time consuming and skills intensive. This slow element of craftsmanship relates well to the slow movement philosophy promulgated by Honore (2004) and Fletcher (2008). Both authors borrow heavily from the Slow Food movement instigated in Italy as a revolt to the penetration of fast food chains, although slow is not literally meant to be the opposite of fast. Decelerating speed is about connecting to real and meaningful situations, such as socializing with friends, participating in local culture, finding time to enjoy a meal, and working to live and not living to work (Honore 2004). Slowness has more to do with a mental disposition this of being unhurried, reflective, calm, patient, and valuing quality as opposed to quantity. Being slow does not mean regressing into a pre-Industrial Revolution utopia. On the contrary, it is more about finding a natural bio-rhythm, this of equilibrium where individuals can balance out the speed of contemporary life while simultaneously remaining slow inside (Fletcher 2008; Honore 2004).

Fletcher (2008) relates the slow philosophy to textiles and fashion by devising a number of fundamental precepts. In the slow fashion concept, the customer becomes a co-producer as he/she participates in the production process. The customer is also conditioned to elect quality over quantity in the sense of consuming less and more responsibly. Companies within the slow fashion mindset engage in fair practices toward their workers across the supply chain, thus drastically improving their livelihoods. They also reduce the use of raw materials by being more resourceful in alternative ways of preserving natural resources. Slow fashion firms use mostly local materials and labor instead of sourcing thousands of miles away from their base. In this way, they reduce their carbon footprint. Moreover, they care about the preservation of local traditional skills. Esthetic considerations are not consumed to the latest fad, but slow firms take a more neo-classic approach to design products with the quality of manufacture, which is certainly more durable. This guarantees that garments can be worn over longer periods of time. Financial viability is secured since slow companies can charge more for their clothing and accessories and customers are willing to pay a premium because they know that fashion products are made in a fair practice ecosystem (Fletcher 2008).

The concept of the relational value chain in clothing production configurations will be discussed in the next section. This is coupled with the theory of the relational view of the firm and dynamic capabilities theories which constitute sources of competitive advantage in networks.

5 Relational Networks in Value Chains

Value chains in textiles and clothing (T&C) production relate to the fragmentation of production and supply operations, a phenomenon which was intensified in advanced economies during the 1990s. In this, retailers and branded manufacturers, usually refer to the buyers, sought to outsource production to lower labor cost countries spanning regions and continents. The so-called buyer-driven value chains pertain to T&C production networks where buyers coordinate production operations by organizing decentralized networks of suppliers. These suppliers in many cases source fabrics, trims, and complete garment assembly in the form of shop-ready merchandise, thus providing full package services to buyers (Gereffi 1999; Gereffi and Memedovic 2003). Within the general taxonomy of the value chains, it is embedded in the concept of the relational chains that the coordination of production can be arranged in networks of variable sizes. These range from a local network which produces specialized clothing products, to networks that span countries and regions (Berger 2006; Gereffi et al. 2005).

In relational networks, collective trust which derives from interdependence prevails, as territorial proximity, reputation of manufacturers, and social coherence are crucial success factors (Berger 2006; Gereffi et al. 2005). Moreover, trust has a number of additional benefits for network members. Continuity of collaborations in adverse market conditions, mutual commitment, and dependencies strengthen the network because all participants acknowledge that benefits will be fairly distributed (Lee and Cavusgil 2006; Poppo and Zenger 2002).

Network theorists adopt the concept of central controllers who have developed abilities for the effective coordination and organization of production activities (Gulati et al. 2000; Jarillo 1988; Shih 2013). Central controllers have the means to obtain and command critical resources in any given network, which helps them to form a structured hierarchy around suppliers (Berger 2006; Shih 2013).

5.1 The Relational View of the Firm and Network Resources

The relational view of a firm derives from the resource-based view theory which posits a fact that a single company is comprised of a certain bundle of critical tangible and intangible resources. These critical resources as they are transformed into capabilities and deployed to the markets constitute possible sources of competitive advantage (Barney 1991; Dyer and Singh 1998; Grant 1991). The relational view puts forward the notion that the company can be viewed not only as an individual entity but also as a nexus of relations. In this context, the company extends its boundaries and forms relationships with other network members by accessing network resources. Thus, competitive advantage can accrue within a network of inter-firm collaborations.

The synergies created by collaborations when companies combine and reconfigure resources can generate specialized capabilities which form a powerful blend of core competences (Barney 1999; Das and Teng 2000; Duschek 2004; Dyer and Singh 1998; Gulati et al. 2000). Core competences relate to the coordination of production processes which in turn correspond to organizational routines (Hamel and Prahalad 1994). Capabilities are generated through these routines which can lead to knowledge (Grant 1991). Knowledge, which resides within individuals in the firm, is capable of strengthening further the capabilities of the firm. In this concept, the firm by forming external collaborations can develop capabilities which are nearly inimitable because they are considered specialized knowledge and as such possible sources of competitive advantage (Conner and Prahalad 1996; Grant 1996; Schroeder et al. 2002).

Another important extension of the resource-based view is the so-called dynamic capabilities where competitive advantage may accrue in changing market conditions. In this context, dynamic capabilities refer to the ability of a company to survive and consequently evolve by reconfiguring resources and regenerating capabilities in order to align itself into changing business circumstances. Top management's resilience can combine internal capabilities and competences and fuse them with external resources. Companies by taking rapid action to respond to these changing market conditions can engender fresh strategies of value creation (Eisenhardt and Martin 2000; O'Regan and Ghobadian 2004).

The following section of the chapter pertains to the selected methodology and methods. The authors attempt to relate the subject presented in the literature, namely the sustainability imperative and the prerequisite of decelerating luxury fashion consumption, to the case study of the tailoring luxury craft. This will provide an understanding of the research objective that competitiveness can coexist within the slowness of luxury artisanal production; a notion that runs contrary to current thinking of meaningless diversification tactics.

6   Research Methodology and Methods

The philosophical view of interpretivism espouses the logic of interpretation and observation in understanding the societal context of the world. This approach is integral to the qualitative research tradition since it ascribes to the idea that knowledge is actively constructed by humans. In this concept, the interpretivist paradigm is focused on the contexts of how humans live and work which helps researchers to understand the cultural and historical background of people in specific societal situations (Ritchie et al. 2013; Creswell 2009). Moreover, researchers' own setting can configure their interpretation because they position themselves in the situational research. This fact assists them to comprehend how their interpretation flows and is also influenced by their own personal, historical, and cultural experiences. Therefore, researchers' intent is to interpret a pattern of meanings which is configured by the actors who shape the subject matter of the research (Ritchie et al. 2013; Creswell 2009).

The principal aim of the qualitative research method is to explain the subjective processes and meanings of how individuals or groups take action or cope in a particular social situation for the qualitative tradition is based on the social construction of reality. This can assist researchers to achieve a holistic overview of the subject matter as they interact with the context of the situation which shapes their investigation (Creswell 2009; Atheide and Johnson 1994).

In order to achieve the objective of this study, the authors make use of descriptions and explanations of secondary data. This is also based on the interpretive context of both authors' working experience in Savile Row and one of the authors' tailoring family background. Moreover, the qualitative approach facilitates the researchers to gain a profound understanding of *how* this enduring competitive advantage can be sustained, through the intricate functions of simultaneous cooperation and competition in the tailoring network (Yin 2009; Denzin and Lincoln 1994). The authors have elected the qualitative approach as more appropriate for this study because the inherent social context of the tailoring network and its' interpretation is difficult to quantify. Therefore, the researchers employ the case study method to explore tailors activities, capabilities, and resources since in most situations they contain relevant information. This may provide an in-depth understanding of the concept of competitiveness within the context of a sustainable luxury craft.

In the next section, the evolutionary modes of survival of the Savile Row tailors are demonstrated. The tailoring techniques are also explained together with the production practices of the network which are instrumental to their competitiveness and consequent growth.

7 Savile Row: A Brief History of a British Institution

Savile Row, the so-called "golden mile of tailoring," is a destination street located between Regent and Bond streets, two of the main commercial arteries in London's West End (Hadley 2006; Norton 2006). The street became the epicenter of men's tailoring businesses in the middle of the nineteenth century with the advent of the Industrial Revolution. Textile merchants sought to sell tailored clothing to the British gentry and for this reason they organized tailoring activities centered on the street and surrounding areas (Anderson 1981). In reality, the esthete Beau Brummel took a very different view from the French dominance of brocade silk jackets which were *de rigueur* under pre-revolution France. He contrived a look that comprised somber colors, namely navy, black, and charcoal gray for the jacket and ankle trousers of which he was the inventor, combined with white shirts and cravats. His inspiration derived from the understated riding attire of the British aristocracy. The founding tailor of the street is accredited to James Poole who in 1846 opened his tailoring establishment to the British aristocracy (Howarth 2003; Savile Row Bespoke Association 2015). The street's tailors in the following decades invented and also produced some of the most remarkable tailored menswear for an array of politicians, actors, celebrities, industrialists, businessmen, and aristocrats from around the world (Hadley 2006; Mellery-Pratt 2014; West 2014).

7.1 Decline and Revival

In the 1980s, during the proliferation of designer ready-to-wear for men, which dominated retail sales worldwide, the Savile Row tailoring tradition was nearly extinct as it was disdained by the younger generation (Conti 2010). The old stuffy image of the premises was compounded by the fact that the business model of the tailoring establishments was confined to an old fogey clientele who looked irrelevant to the sartorial currents of the era (Moreton 2000; O' Cealaigh 2012; Sherwood 2006). The despondency of a bleak future started to change in the middle 1990s, when a new generation of tailors opened premises on the street, introducing novelty of both cuts and colors. Initially, they were not particularly welcomed by the Savile Row establishment as they were considered arrivistes and upstarts, but with the progress of time opinions shifted to the present state of affairs (Conti 2010; Moreton 2000). In more recent time, new names such as Richard James and Ozwald Boateng have appeared; others have been acquired by big clothing corporations thus securing their future and branching out to lucrative international markets. Also a small number of tailor houses have been restored to their former glory by tailoring enthusiasts and investors (Mellery-Pratt 2014; Mills 2011; West 2014).

Currently, Savile Row enjoys a global revival attributed to a number of reasons. These are associated with the fact that the contemporary men are far more interested in how they look, and spend more time and money for their sartorial quests (Mills 2011; Shannon 2013). Additionally, traditional tailoring in general is thought of as an authentic craft made by human hands. This connects to a new notion of luxury and not only to expensive branded clothing (Eisenhammer 2013). Another reason is that tailoring is personal since a suit is made to an individual style and measurements and not to an average body figuration which is the case with ready-to-wear (Norton 2006). Moreover, tailoring can be considered as sustainable luxury for most of the Savile Row establishments making use of materials sourced mainly from British woolen mills and suits produced within their premises (Belcher 2013; Conti 2010; Norton 2006). In addition, a suit that is well tailored in a decent cloth and is not subject to fashion's frivolities can withstand the test of time by serving its wearer for years if not generations. The cutters deliberately leave cloth inlays in the seams for future alterations (Anderson 1981; Belcher 2013; Conti 2010). Some of the tailoring establishments have extended their operations by incorporating ready-to-wear ranges and related accessories, have branched out to international markets, and also have started to advertise to the public by opening stores. These diversification ploys may look like tailors becoming more of a brand; nevertheless, this is essential for their survival. This fact can be related to the famous French houses of Chanel and Hermes and their past and also current practices (Conti 2010; Shannon 2013).

As with any business, the scale of operations is a fundamental prerequisite for growth. However, in a cottage industry like tailoring, economies of scale are not existent no matter how much the business grows its productivity, as human hands have a limited speed capacity. Hence, expansion to off-the-peg ranges and accessories is considered by some tailoring firms a wise growth strategy (Davidson 2012; Mellery-Pratt 2014).

7.2 What Is a Bespoke Tailoring Service?

Bespoke as a tailoring term merits more clarification, for there is confusion with made-to-measure or custom-made. Bespoke derives from the centuries old custom of a gentleman ordering (bespeaking) his suit to his exact specifications in measurements, figuration, style, cloth, and design features, in a unique handmade construction by a master tailor. Made-to-measure retains some of the above characteristics but the fundamental difference is that it makes use of a universal pattern template which is adjusted to the customer's measurements and figuration. Davies (1998) observes that the made-to-measure suits are assembled by a mixed method of mainly sewing machines and hand stitching which is only externally visible. This is for the better and more expensive construction, but for the average made-to-measure suit and shirts, they are all cut and made by laser cutting and sewing machinery which is also reflected in the price.

The tailoring process requires approximately 55 h of a manual labor in order to cut and assemble a two-piece suit (jacket and trousers without a waistcoat). The master tailor or the cutter takes around 35 measurements from the customer and relays this information to an assistant who writes down the measurements in a specific order on a ledger. The master tailor together with the fabric consultant discusses with the customer about his choice of design features, cloth, linings, and buttons. The master tailor then cuts a paper pattern and passes it to a striker who cuts the cloth along the lines of the pattern. In some cases, the master tailor cuts directly onto the cloth. After cutting, the cut pieces are bundled up and handed over to either one tailor who will assemble the whole suit by himself or in bigger establishments to the tailoring workshop where tailors are specialized in different assembly stages. The perfect suit fit requires three or sometimes four fittings at different stages of the construction and the whole process takes more than a month to complete depending on the orders backlog (Anderson 1981; Davidson 2012; Norton 2006).

7.3 The Tailoring Production Network

The physical location of Savile Row and its surrounding areas is of paramount importance to the tailoring businesses because the street name is associated with bespoke services renowned all over the world. There are approximately a total of 116 tailoring businesses comprised of the legendary tailor houses such as Gieves & Hawkes and Henry Poole & Co, to small tailor workshops and sole traders in a variety of business sizes (Hadley 2006).

Tailoring production is organized in a dense network of operations (shown in Fig. 1) where production is divided into two main types. The first type consists of

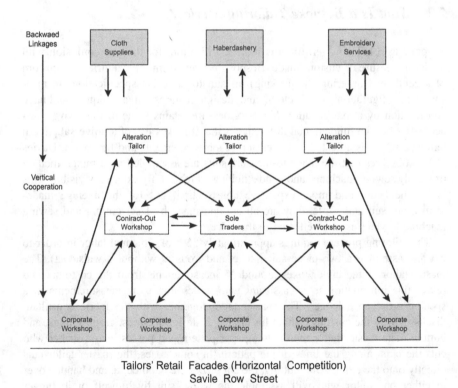

Fig. 1 The relational network of Savile Row (Established by the authors)

the famous tailor houses (17 in absolute numbers) which employ a wide range of staff. They are trained to carry out the complete tailoring process on a division of labor basis, under the corporate workshop. The second type pertains to small workshops comprising a handful of tailors, who either take orders directly from customers and/or receive work from master tailors of the more established houses as in many cases the corporate workshops are overwhelmed with backlog orders. There are also sole trader tailors who work only by taking orders from master tailors, since master tailors only engage in measuring and cutting but do not assemble garments. Moreover, within the network all master tailors cooperate with a number of alteration tailors who are very skillful in garment modifications following instructions from the master tailors after fittings.

The network is also connected to a number of other backward linkages. Cloth merchants can supply suit lengths within a day or two. Haberdashery firms supply among other things, threads, canvases, linings, and buttons. Embroidery firms are involved in embroidery services for monogramming, regimental badges, and also military uniforms. It must be noted that all the above supporting services are within walking distance from any tailor workshop.

The physical proximity and cooperation are very important parameters of all production stages in the network. There is a strong feeling of community among tailors, since a spirit of competition and simultaneously cooperation has been cultivated for decades in the area. Naturally, all tailoring houses compete for orders on the retail side of the business. However, collaboration is the main strength of the network operation as in many cases tailors do favors to one another. A tailor may send a customer to a fellow tailor if he believes that a customer's wishes do not befit his style of tailoring. Thus, competition is horizontal in retail operations and cooperation is vertical in a relational mode of production (Hadley 2006).

8 Concluding Comments and Perspectives

The contemporary notion of luxury seems to embrace once more its perennial notions of superior craftsmanship, timelessness in design, and durability as customers seek more meaningful ways to engage with luxury brands and their product ranges. This reversal to authenticity and original values has also been influenced by the fact that the sector was reprehensible in sustainability terms while it was expected to behave in a more responsible manner within the context of sustainable development. In order to remedy the previous state, initiatives are undertaken by some luxury conglomerates and they seem to head toward the right direction although there is a long distance to cover toward a truly sustainable luxury ecosystem. It must also be noted that these initiatives constitute sincere intentions which are not solely driven by ruthless business practices, but they take the environment and other members of the supply chain into consideration.

This chapter has demonstrated in the case study that competitiveness can be achieved in at least a section of the luxury sector as this is the case with the Savile

Row tailors network. The tailoring craft also relates well with the fundamental slow fashion philosophy precepts which seem to take center stage in the discourse on sustainable development in fashion. More analytically, competitiveness and slow fashion are connected to the tailoring network for the following major reasons.

The relational value chains and networks can generate sources of competitive advantage which is the case with Savile Row tailoring operations. The central controllers namely the established tailor houses capture the vital resources of the network and form a loose hierarchy with a number of participants who conform to the specific house style. Despite the looseness in the hierarchy, this has solid foundations because it is based on trust and the collective perception that financial rewards are fairly distributed. Simultaneously, the backward linkages procure all necessary supplies for the efficient and smooth operation of the network. Apprenticeships are also provided by the majority of the bigger establishments in line with the slow fashion principles of preserving the craft traditions, thus reinforcing further the network and securing its future.

Another important parameter inherent in the network is dynamic capabilities as new generations of tailors have transformed the old fogey perceptions of bespoke tailoring into a contemporary and more relevant artisanal image cherished by initiated customers around the world. Moreover, new sources of competitive advantage have been generated as some astute owners/managers have realized that there are opportunities to be reaped by on the one hand preserving a famous craft and on the other transforming the tailor house into more of a brand by introducing well-tailored off-the-rack ranges and other accessories. Dynamic capabilities also relate to some of the slow fashion principles, such as timelessness, where all tailor houses engage into a style which despite its multiple variations remains a classic within the context of the quintessential London cut. The sculptural qualities of the London cut itself has become a global ambassador for the Savile Row tailoring techniques as it is renowned by luxury *connoisseurs*. In this concept, specialized knowledge which is generated by tailors' specific capabilities in a tailoring style puts forward the possibility of competitive advantage, since there are not many others who can imitate a very particular and unique style. Even if competitors ever manage to get close to a specific style, this will be considered as inauthentic by Savile Row purists and soon will be disdained.

The slow fashion philosophy of the co-producer element inherent in the process is represented well in Savile Row because customers co-design their suits with the master tailor. Since a suit requires three to four fittings, the tailor literally during the fittings sculpts the suit to the customer figuration. The master tailors are also experts in hiding small body deformities by tweaking the patterns and molding the cloth so that the end result looks remarkable. In this sense, since the customer participates in the fitting process and becomes a co-producer, this clothed body serves as the model for tailoring.

Moreover, the slow fashion principle of buy less but buy better is associated with Savile Row because usually customers do not engage in wasteful consumption patterns. They buy their suits which they keep for decades not to mention that in cases they hand them down to their sons following the British tradition. Even if they do not hand their suits down to the following generation, the maintenance of the suits is always guaranteed by the tailor houses, such as letting out or taking in seams because all bespoke suits are cut with extra cloth inlays on the seams.

Durability, which forms another slow fashion principle, is also maintained since the overwhelming majority of the suits are made with exceptional quality worsted high twist wools by mainly British mills which are durable by definition. This is also complemented by the fact of superior stitching techniques which means that the overall suit shape ages gracefully.

This leads to another slow fashion precept, the use of local materials. Indeed, in Savile Row, the majority of cloths and trims are either British or European made, thus they do not significantly burden the environment in long haul transportation. They also maintain employment of the sensitive textiles and clothing sectors within European borders.

On balance, the authors can analytically verify that sustainable luxury and competitiveness can coexist beyond huge diversification projects of mass luxury product ranges which have ushered the luxury sector into wasteful patterns of production and meaningless consumption. The guiding principle of sustainable development must permeate all strategic decisions and operations at the corporate level, if luxury customers are to continue relishing unique handmade goods in the long term.

References

Agins T (2000) The end of fashion: how marketing changed the clothing business forever. Harper Collins Publishers, New York

Anderson SH (1981) The tailors of Savile Row. The New York Times, New York. http://www.nytimes.com/1981/02/08/business/the-tailors-of-savile-row.html. Accessed 25 Oct 2014

Aristotle (1998) The Nicomachean ethics. Oxford University Press, Oxford

Atheide DL, Johnson JM (1994) Criteria for assessing interpretive validity in qualitative research. In: Denzin NK, Lincoln YS (eds) Handbook of qualitative research. SAGE Publications, Thousand Oaks, pp 485–499

Atwal G, Williams A (2009) Luxury brand marketing- the experience is everything. J Brand Manag 16:338–346

Barney JB (1991) Firm resources and sustained competitive advantages. J Manag 17:99–120

Barney JB (1999) How a firm's capabilities affect boundary decisions. Sloan Manag Rev 137–145

Belcher D (2013) Bespoke, and built to last, in the finest Savile style. International New York Times, New York. http://www.nytimes.com/2013/11/19/fashion/bespoke-suits-built-to-last.html?_r=0. Accessed 23 Nov 2013

Bellaiche JM, Mei-Pochtler A, Hanisch D (2010) The new world of luxury. The Boston Consulting Group, Boston. https://www.bcg.com/documents/file67444.pdf. Accessed 12 May 2014

Bendell J (2012) Elegant disruption: how luxury and society can change each other for good. Asia Pacific work in progress research papers series, vol 9. Griffith University, South Brisbane

Bendell J, Kleanthous A (2007) Deeper luxury. WWF-UK, London. http://www.wwf.org.uk/deeperluxury/. Accessed 4 Sept 2014

Berger S (2006) How we compete. Doubleday of Random House Inc., New York

Berry CJ (1994) The idea of luxury: a conceptual and historical investigation. Cambridge University Press, Cambridge

Berthon P, Pitt L, Parent M, Berthon JP (2009) Aesthetics and ephemerality: observing and preserving. Calif Manag Rev 52:45–66

BOF Team (2014) The year fashion woke up. Business of fashion Ltd., London. http://www.businessoffashion.com/2014/12/year-fashion-woke.html. Accessed 8 Jan 2015

Brundtland GH (1987) Our common future, chairman's forward. Report of the world Commission on environment and development. United Nations, Oslo. http://www.un-documents.net/ocf-cf.htm. Accessed 10 Oct 2013

Cachon GP, Swinney R (2011) The value of fast fashion: quick response, enhanced design and strategic consumer behavior. Manage Sci 57:778–795

Clifford S (2013) Some retailers say more about their clothing's origins. The New York Times, New York, 8 May 2013

Cline EL (2013) Overdressed: the shockingly high cost of cheap fashion. Porfolio. Penguin Books, New York

Conner KR, Prahalad CK (1996) A resource-based theory of the firm: knowledge versus opportunism. Organ Sci 7:477–501

Conti S (2010) Reinventing Savile Row. Women's Wear Daily, New York, 18 Oct 2010

Creswell JW (2009) Research design. Qualitative, quantitative and mixed methods approaches, 3rd edn. SAGE Publications, Thousand Oaks

Danziger P (2005) Let them eat cake! Marketing luxury to the masses, as well as the classes. Deerborn Trade Publishing, Chicago

Das TK, Teng BS (2000) A resource-based theory of strategic alliances. J Manag 26:31–61

Daveu MC (2014) Placing sustainability at the heart of Kering. Business of fashion Ltd., London. http://www.businessoffashion.com/2014/10/placing-sustainability-heart-kering.html. Accessed 1 Oct 2014

Davidson A (2012) What's a $4,000 suit worth? The New York Times, New York. http://www.nytimes.com/2012/09/09/magazine/whats-a-4000-suit-worth.html. Accessed 4 Sept 2013

Davies H (1998) Fast track: suits and blazers cut by lasers. The Independent, London. http://www.independent.co.uk/arts-entertainment/fast-track-suits-and-blazers-cut-by-lasers-1146935.html. Accessed 10 Oct 2014

De Botton A (2000) The consolations of philosophy. Vintage International, New York

Denzin NK, Lincoln YS (1994) Entering the field of qualitative research. Handbook of qualitative research. SAGE Publications, Thousand Oaks, pp 1–17

Doran S (2013) The new luxury is luxury for all, suggests Jean-Noel Kapferer. Luxury Society. http://luxurysociety.com/articles/2013/04/the-new-luxury-is-luxury-for-all-suggests-jean-noel-kapferer. Accessed 10 Oct 2014

Dubois B, Czellar S, Laurent G (2005) Consumer segments based on attitudes towards luxury: empirical evidence from twenty countries. Market Lett 16:115–128

Duschek S (2004) Inter-firm resources and sustained competitive advantage. Manag Revue 15:53–73

Dyer JH, Singh H (1998) The relational view: cooperative strategy and sources of interorganizational competitive advantage. Acad Manag Rev 23:660–679

Eisenhammer S (2013) Timeless suits from London's Savile Row back in fashion. Thomson Reuters. http://uk.reuters.com/article/2013/02/20/uk-fashion-menswear-savilerow-idUKBRE91J0IY20130220. Accessed 18 July 2014

Eisenhardt KM, Martin JA (2000) Dynamic capabilities: what are they? Strateg Manag J 21:1105–1121

Fletcher K (2008) Sustainable fashion and textiles: design journeys. Earthscan, London

Frank RH (2014) Conspicuous consumption? Yes, but It's not crazy. The New York Times, New York. http://www.nytimes.com/2014/11/23/upshot/conspicuous-consumption-yes-but-its-not-crazy.html?abt=0002&abg=1. Accessed 22 Dec 2014

Friedman, V (2010) Sustainable fashion: what does green mean? Financial Times, London. http://www.ft.com/cms/s/2/2b27447e-11e4-11df-b6e3-00144feab49a.html. Accessed 15 May 2013

Gam HJ, Banning J (2011) Addressing sustainable apparel design challenges with problem-based learning. Cloth Text Res J 29:202–215

Gardetti MA, Torres AL (2013) Introduction. J Corp Citiz 5–8

Gereffi G (1999) International trade and industrial upgrading in the apparel commodity chain. J Int Econ 48:37–70

Gereffi G, Memedovic O (2003) The global apparel value chain: what prospects for upgrading by developing countries. United Nations, Vienna

Gereffi G, Humphrey J, Sturgeon T (2005) The governance of global value chains. Rev Int Polit Econ 12:78–104

Grant RM (1991) The resource-based theory of competitive advantage: implications for strategy formulation. Calif Manag Rev 33:114–135

Grant RM (1996) Toward a knowledge-based theory of the firm. Strateg Manag J 17:109–122

Guercini S, Ranfagni S (2013) Sustainability and luxury: the Italian case of a supply chain based on native wools. J Corp Citiz 52:76–89

Gulati R, Nohria N, Zaheer A (2000) Strategic networks. Strateg Manag J 21:203–215

Hadley G (2006) Bespoke tailoring in London's West End. Planning & City Development, London. http://transact.westminster.gov.uk/docstores/publications_store/Bespoke_tailoring_report_March_2006.pdf. Accessed 1 Nov 2015

Halzack S (2014) The definition of luxury retail is being 'shattered'. The Washington Post, Washington DC. http://www.washingtonpost.com/news/business/wp/2015/01/12/the-definition-of-luxury-retail-is-being-shattered/. Accessed 12 Feb 2015

Hamel G, Prahalad CK (1994) Competing for the future. Harvard Business School Press, Boston

Hilton M (2009) The legacy of luxury: moralities of consumption since the 18th century. J Cons Culture 4:101–123

Honore C (2004) In praise of slowness, challenging the cult of speed. Happer Collins Publishers, New York

Howarth S (2003) Henry Poole: founders of Savile Row: the making of a legend. Bene Factum Publishing Ltd., Honiton

Husic M, Cicic M (2009) Luxury consumption factors. J Fash Market Manag Int J 13:231–245

Jackson F (2011) Giving the environment a sporting chance. The Times, London. http://www.sportanddev.org/?1072/Giving-the-Environment-a-Sporting-Chance. Accessed 6 July 2014

Jarillo JC (1988) On strategic networks. Strateg Manag J 9:31–41

Joy A, Sherry JF, Venkatesh A, Wang J, Chan R (2012) Fast fashion, sustainability, and the ethical appeal of luxury brands. Fash Theory J Dress Body Cult 16:273–296

Kahn J (2009) Luxury-goods makers embrace sustainability. The New York Times, New York. http://www.nytimes.com/2009/03/27/business/worldbusiness/27iht-sustain.html. Accessed 27 Sept 2014

Kapferer JN (2004) The new strategic brand management: advanced insights and strategic thinking. Free Press, New York

Kapferer JN (2010) All that glitters is not green: the challenge of sustainable luxury. Eur Bus Rev 40–45

Kapferer JN (2012) Why luxury should not delocalize: a critique of a growing tendency. Eur Bus Rev 56–61

Kapferer JN, Bastien V (2012) The luxury strategy. Kogan Page, London

Kemp S (1998) Perceiving luxury and necessity. J Econ Psychol 9:591–606

Kiron D, Kruschwitz N, Haanaes K, Reeves M, Fuisz-Kehrbach SK, Kell G (2015) Joining forces: collaboration and leadership for sustainability. MIT Sloan Manag Rev. http://sloanreview.mit.edu/projects/joining-forces/. Accessed 5 Feb 2015

Kleanthous A (2011) Simply the best is no longer simple. The Times, London. http://theraconteur.co.uk/category/sustainability/sustainable-luxury. Accessed 6 July 2014

Lee Y, Cavusgil ST (2006) Enhancing alliance performance: the effects of contracted-based versus relational-based governance. J Bus Res 59:896–905

Mellery-Pratt R (2014) A row of opportunity, part 1. Business of Fashion. http://www.businessoffashion.com/2014/01/row-opportunity-part-1.html. Accessed 10 March 2014

Mills S (2011) The tailors putting Savile Row back on the map. Evening Standard, London. http://www.standard.co.uk/lifestyle/the-tailors-putting-savile-row-back-on-the-map-6576484.html. Accessed 11 March 2015

Moreton C (2000) It's scissors at dawn in Savile Row. The Independent. http://www.independent.co.uk/news/uk/this-britain/its-scissors-at-dawn-in-savile-row-710922.html. London. Accessed 12 Jan 2015

Norton K (2006) Savile row never goes out of style. Bloomberg Businessweek, New York

Nueno JL, Quelch JA (1998) The mass marketing of luxury. Bus Horiz 41:61–68

O' Cealaigh J (2012) London: the tailors of Savile Row. The Daily Telegraph, London. http://www.telegraph.co.uk/travel/destinations/europe/uk/london/9333705/London-the-tailors-of-Savile-Row.html. Accessed 15 Jun 2014

O'Regan N, Ghobadian A (2004) The importance of capabilities for strategic direction and performance. Manag Decis 42:292–312

Phau I, Pendergast G (2000) Consuming luxury brands: the relevance of the 'rarity principle'. Brand Manag 8:122–138

Pine BJ, Gilmore JH (1999) The experience economy: work is theatre and every business a stage. Harvard Business Press, Boston

Poppo L, Zenger T (2002) Do formal contracts and relational governance function as substitutes or complements? Strateg Manag J 23:707–725

Postrel VI (2003) The substance of style: how the rise of aesthetic value is remaking commerce, culture, and consciousness. HarperCollins, New York

Rad A (2012) Luxury brand exclusivity strategies—an illustration of a cultural collaboration. J Bus Admin Res 1:106–110

Rambourg E (2014) The bling dynasty: why the reign of Chinese shoppers has only just begun. Wiley, Singapore

Ritchie J, Lewis J, McNaughton- Nichols C, Ormston R (2013) Qualitative research practice: a guide for social science students and researchers. SAGE Publications, Los Angeles

Rose M (2014) Inspections roil garment industry in Bangladesh. Taipei Times, Taipei. http://www.taipeitimes.com/News/editorials/archives/2014/06/30/2003593986. Accessed 30 Sept 2014

Savile Row Bespoke Association (2015) Savile Row Bespoke. http://www.savilerowbespoke.com/. Accessed 25 Oct 2014

Schroeder RG, Bates KA, Junttila MA (2002) A resource-based view of manufacturing strategy and the relationship to manufacturing performance. Strateg Manag J 23:105–117

Shannon S (2013) London's Savile Row tailors strive to stay a cut above. Bloomberg Businessweek, New York. http://www.bloomberg.com/bw/articles/2013-08-22/londons-savile-row-tailors-strive-to-stay-a-cut-above. Accessed 10 Oct 2014

Sherwood J (2006) Big enough for the both of us? Financial Times, London. http://www.ft.com/intl/cms/s/0/0d87dc8a-1e9f-11db-9877-0000779e2340.html?siteedition=intl#axzz3Z4hFOR6L. Accessed 2 Mar 2015

Shih WYC (2013) Investigation of the competitiveness of a textile and apparel manufacturer: a case study in Taiwan. PhD thesis, School of Materials, University of Manchester, Manchester, UK

Siegle L (2011) Why fast fashion is slow death for the planet. Why fast fashion is slow death for the planet. http://www.theguardian.com/lifeandstyle/2011/may/08/fast-fashion-death-for-planet. The Guardian: Fashion The Observer, London, Accessed 15 Mar\2015

Silverstein MJ, Fiske N (2003) Luxury for the masses. Harvard Bus Rev 81:48–57

Simmel G (1957) Fashion. Am J Sociol 62:541–558

Smitha E (2011) Screwing mother earth for profit. Watkins Publishing, London

Stankeviciute R, Hoffmann J (2010) The impact of brand extension on the parent luxury fashion brand: the cases of Giorgio Armani, Calvin Klein and Jimmy Choo. J Global Fash Market 1:119–128

Thomas D (2007) Deluxe: how luxury lost its luster. The Penguin Press, New York

Transparency Market Research (2014) Global luxury goods market—industry analysis, size, share, growth, trends, and forecast 2014–2020. http://www.transparencymarketresearch.com/luxury-goods-market.html. Accessed 11 Nov 2014

Truong Y, Simmons G, McColl R, Kitchen PJ (2008) Status and conspicuousness—are they related? Strategic marketing implications for luxury brands. J Strateg Market 16:189–203

Tsouna V (2007) The epistemology of the Cyreanaic school. Cambridge University Press, Cambridge

Tungate M (2012) Fashion brands: branding style from Armani to Zara. Kogan Page, New York

Twitchell JB (2002) Living it up: America's love affair with luxury. Columbia University Press, New York

United Nations (2012) Back to our common future: sustainable development in the 21st century. Department of Economic and Social Affairs. United Nations, New York. https://sustainabledevelopment.un.org/content/documents/UN-DESA_Back_Common_Future_En.pdf. Accessed 24 Aug 2014

United Nations Environment Programme (2014) Integrated governance: a new model of governance for sustainability. UNEP Finance Initiative, Geneva. http://www.unepfi.org/fileadmin/documents/UNEPFI_IntegratedGovernance.pdf. Accessed 19 Dec 2014

United Nations Global Compact Office (2014) A guide to traceability: a practical approach to advance sustainability in global supply chains. United Nations, New York. http://www.bsr.org/reports/BSR_UNGC_Guide_to_Traceability.pdf. Accessed 19 Dec 2014

Veblen T (1994) Theory of the leisure class. Penguin, New York

Vickers JS, Renand F (2003) The marketing of luxury goods: an exploratory study- three conceptual dimensions. Market Rev 3:459–478

Vigneron F, Johnson LW (2004) Measuring perceptions of brand luxury. J Brand Manag 11:484–506

Weber C (2007) The devil sells Prada. The New York Times, New York. http://www.nytimes.com/2007/08/26/books/review/Weber-t.html?pagewanted=all. Accessed 26 Aug 2014

West K (2014) Savile Row tailor fears overseas threat to rich tapestry of tradition. The Guardian, London. Savile Row tailor fears overseas threat to rich tapestry of tradition. Accessed 14 Nov 2014

Wiedmann KP, Hennigs N, Siebels A (2007) Measuring consumers' luxury value perceptions: a cross-cultural framework. Acad Market Sci Rev 7:333–361

Woolf R (2010) Pleasure and desire. Cambridge University Press, Cambridge

Yin R (2009) Case study research: design and methods, 4th edn. SAGE Publications, Los Angeles

Zilioli U (2014) The Cyrenaics. Routledge Publishers, Oxon

Printed in the United States
By Bookmasters